Approach your exams the IB way

Mathematical studies SL

IB DIPLOMA PROGRAMME

Merryn Dawborn-Gundlach
Jane Forrest
Nadia Stoyanova Kennedy
Paula Waldman de Tokman

**IB Prepared
Approach your exams the IB way
Mathematical studies SL**

Published March 2009

International Baccalaureate
Peterson House, Malthouse Avenue, Cardiff Gate
Cardiff, Wales GB CF23 8GL
United Kingdom
Phone: +44 29 2054 7777
Fax: +44 29 2054 7778
Website: http://www.ibo.org

The International Baccalaureate (IB) offers three high quality and challenging educational programmes for a worldwide community of schools, aiming to create a better, more peaceful world.

© International Baccalaureate Organization 2009

The rights of Merryn Dawborn-Gundlach, Jane Forrest, Nadia Stoyanova Kennedy and Paula Waldman de Tokman to be identified as authors of this work have been asserted by them in accordance with sections 77 and 78 of the Copyright, Designs and Patents Act 1988.

All rights reserved. No part of this publication may be reproduced, stored in a retrieval system, transmitted in any form or by any means, electronic, mechanical, photocopying, recording, or otherwise, without the prior permission of the publishers.

The IB is grateful for permission to reproduce and/or translate any copyright material used in this publication. Acknowledgments are included, where appropriate, and, if notified, the IB will be pleased to rectify any errors or omissions at the earliest opportunity.

IB merchandise and publications can be purchased through the IB store at http://store.ibo.org. General ordering queries should be directed to the sales and marketing department in Cardiff.

Phone: +44 29 2054 7746
Fax: +44 29 2054 7779
Email: sales@ibo.org

British Library Cataloguing in Publication Data.
A catalogue record for this book is available from the British Library.

ISBN: 978-1-906345-14-3

Cover design by Pentacor**big**
Typeset by Wearset Ltd
Printed and bound in Spain by Edelvives

Item code 5025

2013 2012 2011 2010 2009
10 9 8 7 6 5 4 3 2 1

Acknowledgments
Helen Thomas and Peter Blythe for advice on IB mathematical studies standard level

Table of contents

Chapter 1	Introduction	1
Chapter 2	Get to know your exam paper	3
Chapter 3	Command the command terms	5
Chapter 4	Top tips	7
Chapter 5	Number and algebra	9
Chapter 6	Sets, logic and probability	25
Chapter 7	Functions	43
Chapter 8	Geometry and trigonometry	64
Chapter 9	Statistics	80
Chapter 10	Introductory differential calculus	103
Chapter 11	Financial mathematics	122
Chapter 12	Are you ready?	135

1. Introduction

As an IB student, you are provided with many resources on the path to your final mathematical studies standard level (SL) exams. This book is just one of the resources to help you prepare.

How to use this book

This book contains lots of advice and information to be used throughout the year, but it cannot replace any of the content material that you are expected to have learned before taking your exams.

The main function of this book is to help you to prepare for your exams. It includes useful advice on how to approach exam questions. Real IB student answers are accompanied by examiner commentary, highlighting how marks were gained or lost. We suggest that before you read the students' responses, you try to answer the question by yourself.

What is in this book?

- **Chapter 2** focuses on the structure of the two exam papers: paper 1 and paper 2. We **break down** both the papers to help you to understand what you will find in an exam paper, and what is expected on each of the papers.

- **Chapter 3** explains a selection of the most important **command terms**. These are key words used in exam questions that tell you what is expected in your written answers.

- **Chapter 4** summarizes the main recommendations that teachers from all around the world have identified as **key advice** for IB mathematical studies SL students.

- **Chapters 5 to 11** have been broken up to match the **syllabus topics** you are taught. We have omitted "Introduction to the graphic display calculator", as this topic has only been allocated 3 hours' teaching time, and is intended to give your teacher time to show you the main buttons on your graphic display calculator (GDC). However, we have made GDC references throughout the book.
  - The first half of each chapter reviews **concepts** that you will have learned or will be learning during your studies. Examples will be used to highlight how these concepts may look in exam questions, along with some "**Be prepared**" examiner help. Of course, it would be impossible to include every single piece of syllabus content in this book. So, when you are preparing for your exams, you must also use your own class notes and relevant texts.
  - The second half of each chapter prepares you for the **types of questions** you will meet in your exams. Questions have been selected on their relevance to the syllabus topic. This results in a mixture of past questions from different parts of the exam papers. For each question, we provide three answers, which represent low-, medium- and high-quality answers. These are actual answers, written by IB students like you. The exam questions are accompanied by suggestions to help you to answer them. Please be aware that these suggestions are not the only way to approach the question, and should only be used as guidance. To help

you to gain an insight into the way an examiner works, for each answer we provide you with the mark that was given and an explanation, written by an IB examiner, of key areas where marks were lost and gained. At the end of the set of answers, we provide an examiner report. This is to ensure that you are fully aware of common errors and ways to avoid them.

- In **chapter 12**, you will find a full set of exam papers. These allow you to see exactly what to expect on the big day and to **put into practice** what you have learned from this book.

Please remember that simply reading this book will not guarantee you a high mark. It is a good understanding of mathematical knowledge and how it can be applied during the exam that will help you to achieve success.

We hope that you find the rest of the book useful, and we wish you good luck in your mathematical studies SL exams.

2. Get to know your exam paper

This chapter will help you to understand the main features of the exam papers and what to look out for. There are some points that are common to both papers, but others are specific to individual papers.

General

- Each paper is 90 minutes long, and each paper is worth 40%. The final 20% is your project mark.

- In these papers, there are 90 marks in 90 minutes, so you need to manage your time well. Before the exam starts, you are allowed 5 minutes reading time, during which you are **not** allowed to write or use your GDC. Use this time wisely. Decide which questions you feel most comfortable with and plan to answer those first. The marks allocated for each part give an indication of the amount of work required, and the number of steps needed to solve the particular question.

- You need to know all the topics in the syllabus. There is no point choosing your favourite topics and hoping the others won't come up—they will! The paper writers try their best to include as many topics as possible—but they cannot ask questions on every topic or the papers would be much longer. If a certain topic is not tested in paper 1, then it will probably appear in paper 2. So if geometry and trigonometry has not appeared in paper 1, make sure you revise it well.

The following are ways in which 1 mark is easily lost:
- **accuracy penalty (AP)**—give your answers either exactly or to three significant figures (unless another level of accuracy is asked for in the question)
- **unit penalty (UP)**—include all units of measurement
- **financial penalty (FP)**—answer financial questions to the degree of accuracy given in the question.

They may only apply to 1 mark per paper for each type of mark, but potentially that is 3 marks lost for each paper. They all add up, and could make a difference to your final grade.

Later on in the book you will come across other abbreviations that you may not be familiar with. The table that follows will help you to understand what these mean.

M	Marks awarded for **Method**. Always show working out because some marks are awarded for the method and some for the answer.
A	Marks awarded for an **Answer** or for **Accuracy**. Make sure that you give your answer either exactly or to three significant figures or to the accuracy asked for in the question.
C	Marks awarded for **Correct** answers (irrespective of working shown). This is only true in paper 1. In paper 2 you must show your working out unless it is a "write down" question.

R — Marks awarded for clear **Reasoning**. If you are asked to explain your answer, then give a mathematical explanation.

ft — Marks that can be awarded as **follow-through** from previous results in the question. So, if your answer to one part of a question is wrong but you have used this answer correctly in the following part, then you will be awarded the marks for it.

G — Marks awarded for using your **calculator** correctly. Remember to show what you are doing unless it is a "write down" question.

AG — No marks are awarded. It stands for **Answer given** and is used in "show that" questions.

Paper 1

- This is made up of 15 short questions, each worth 6 marks.

- Questions can be taken from anywhere in the syllabus—but not every topic will be examined in paper 1.

- Questions range from easy to hard, with the easier questions usually at the beginning rather than at the end. This means that you can spend a little bit longer on the harder questions. However, we all like different topics. Decide which questions you feel you can comfortably answer and do those ones first, regardless of their difficulty level.

- Several questions have two, three or four parts that are connected. If you cannot answer some parts, then leave them and go on to the next question. You are more likely to get marks for the first part of one question than for the last part of the previous question.

- Answer the questions in the question book. You can attach extra sheets. Put your working in the box. Don't worry if it spills outside the box. There is no "going outside the box penalty"! Transfer your answers carefully to the answer space.

- Always show all your working. In paper 1 you will be awarded full marks if your answer is correct and you have not shown any working out. However, it is always safer to show your working. A wrong answer with no working receives no marks, whereas a wrong answer with some correct working will receive method marks.

Paper 2

- This is made up of five long questions.

- Questions range from easy to hard, and can be taken from the whole syllabus—but not every topic will be examined in paper 2.

- Some questions are split into two separate parts that are not always related, so beware of changes in the syllabus area.

- Do not write any answers or fill in tables on the question paper—this does not get sent to the examiner. Write everything on the answer sheets provided (as these do get sent to the examiner). Do not mix up parts of one question with parts of another.

- Start your answer to each question on a new page. If you cross out anything, then the examiner will not mark it—even if it is correct! Only cross out work if you have written something else to replace it.

- It is very important to show your working in paper 2. You only get method marks and follow-through marks if your working is shown. If your answer is correct but you have not shown any working out, then you may not be awarded full marks.

- If you have made a mistake in one part of the question and use this wrong answer for the next parts, then, if you show your working out, you will be awarded follow-through marks. If you show no working out, then you get no marks. Don't be lazy—write down your working out. The examiner cannot mark what you did in your head or on your GDC.

- In paper 2 it is important that you can give a reason for your answer. For example, state why you are accepting or rejecting the null hypothesis; elaborate on what you can say about the correlation coefficient.

- Put labels and scales on the axes of your graphs. This will guarantee you 1 mark. Always use the scales given in the question, as these are usually the ones that fit the graph best.

3. Command the command terms

In this chapter you will gain an understanding of command terms, often referred to as key terms, and how they relate to exam questions.

Command terms are the main words you should be looking for in the exam question(s). They describe the type of action you are asked to take in order to arrive at the answer to a given question. In some cases, you may be asked to provide the answer only. In other cases, you may be asked to illustrate the entire process step by step. It is important that you can distinguish between the different instructions, so that you can determine how much depth you should provide in an answer.

Below is a list of command terms used frequently in exams, followed by some useful definitions. Be aware that this list is not complete, so do not worry if you come across a term that is not in the list. You can always check its definition with your teacher. Take time to familiarize yourself with them and practise their definitions when you come across them later in the book. Remember, you can always highlight the command term (or any other important words) on the exam paper to help you throughout the question. To help you with this, we have circled the command terms within the exam questions.

Calculate **Find** **Determine**	This means that you **do** have to show your work or working out. These terms are used when you need to do some working in order to arrive at the answer. So, if you want full marks, do not forget to write down all your steps. For example: • If you are using graphs on your GDC, draw a **sketch** of the graphs you are using. • If you are using a cosine rule, **write down** the cosine rule with the correct numbers inserted into the correct places and then you can write down your answer.
Differentiate	Find the derivative of a given function. For example: • If $y = 2x^3 + 5$ then $\frac{dy}{dx} = 6x^2$ Be careful with negative powers! Remember that $-4 - 1 = -5$ So, if $f(x) = 3x^{-4} + 5x$ then $f'(x) = -12x^{-5} + 5$
Draw	You must draw an **accurate** graph or diagram. This is the time to be neat and precise. For example, the following are examples of ways in which you would have to be accurate: • x and y intercepts, if any, must be marked • turning points must be in the correct place • asymptotes should be drawn in with a dotted line.

Factorize	This means that you have to write the expression as a product of at least two factors. For example: • a difference of two squares, $x^2 - 4 = x^2 - 2^2 = (x-2)(x+2)$ • a trinomial, $x^2 + 3x + 2 = (x+1)(x+2)$
Hence	This is a big hint. Use your previous results from calculations already performed for the same question to work out the answer or to show that the statement following "hence" is true. For example: • (a) Write $\frac{3}{x^2}$ in the form $3x^a$ where $a \in \mathbb{Z}$. (b) Hence, differentiate $y = \frac{3}{x^2}$, giving your answer in the form $\frac{b}{x^c}$, where $c \in \mathbb{Z}^+$.
Hence or otherwise	You do not have to use your previous answer. Your answer to the previous part of the same question is usually the most useful. But if you know another method that works, then that is okay too.
Justify	Give a valid reason for your answer. Just saying "it is correct" or "yes" or "no" will not get you any marks. For example: • you accept the null hypothesis because the chi-squared value is less than the critical value or the p-value is greater than the level of significance • you cannot use a regression line to predict the outcome of using a certain value if that value is outside the given range.
Plot	Mark points on a diagram, checking the scale carefully to avoid making mistakes.
Show that	Show all the working out needed in order to get the answer given. Examiners usually use "show that" when you need the answer to continue with the rest of the question. If you obtain another answer, then forget it and use the one that is given. Don't work backwards from the answer. Students still do this, and it does not usually work well! Use the answer given in the question for the remaining parts. If you do not, you will only get the method marks and the marks for the correct answer will be lost.
Sketch	Do not spend all your time perfecting a masterpiece, as a sketch of the graph does not have to be as accurate as when you are drawing the graph. The important fact here is to show the shape of the function. You must show some indication of the scale on your axes. The following points should be in approximately the correct place: • axes intercepts • turning points (if there are any) • asymptotes (if there are any).
Solve	Find the solution usually in the form $x = \ldots$ and show step by step how you arrived at the solution. This will help you to avoid making mistakes. For example: • Solve the equation $3(2x+1) = x + 8$ $\Rightarrow 6x + 3 = x + 8 \Rightarrow 5x = 5 \Rightarrow x = 1$
Write down	This term is used in a question whose answer is usually arrived at as a result of a fairly simple mathematical operation. This means you **do not** have to show any working. You will get full marks for just writing down the answer. So, do not waste time trying to think of some working out to write down.

4. Top tips

Here are some extra tips that will help you throughout your exams. Some of them may be obvious, but these are sometimes the ones that students forget!

Before the exam

Be aware of what you can bring into the exam room

- You can only bring in what you need in order to sit the exam. This is a pen, pencil, eraser, sharpener, ruler and GDC (plus a scientific calculator if you want). You are not allowed to bring in the instruction booklet for the calculator.

- If your first language is not English, then you are allowed a simple translating dictionary for your first language.

- You may bring water into the exam room, but no other drinks or food!

- Your equipment must be in a see-through bag.

Always check your GDC

- Make sure your **calculator** is in degree mode and has new batteries. You want to avoid your final answer disappearing just before you write it down. (You are allowed to take spare batteries.)

- When **entering data**, be careful not to make mistakes. Do not use calculator notation in your answer. For example, 3.2E-3 is not acceptable notation but 3.2×10^{-3} is correct mathematical notation. Use your calculator well—remember that it has brackets too!

Double check

- Check that you know **IB notation**. You need to know the notation that will appear in your mathematical studies SL exam. However, you may use other notation when writing your answers, provided that the notation you use is widely recognized. For example, IB notation is $3 \leq x < 5$, but other acceptable forms are [3, 5) or [3, 5[.

- Check that you are familiar with the **information booklet**. Do not waste time looking through the booklet to see if a formula is there. You should know which ones are there and where to find them. When copying the formula, do not rush and write it down incorrectly, as you will get no marks.

- Check that you are familiar with the **command terms**. We have given you these in chapter 3.

During the exam

Watch the clock

- Good time keeping is essential—as a rough guide use 1 mark for 1 minute!

- Some questions will be easier than others and will take less time.

- Start with the questions that you find easier. This will leave you more time to do the questions that you find harder and will build up your confidence.

- The worst thing you can do is panic. If you find yourself starting to panic, take a deep breath and find a nice easy question to relax and settle you.

- At the end of the exam, if you do have time left over, recheck your work and final answers.

Make your graphs a work of art!

- Use a ruler for straight lines and draw graphs in pencil. You never know when you may need to erase something.

- Put labels and scales on your graphs.

- Use the scales given in the question, and make sure that your scales are evenly spaced.

- If you have to plot points, then take your time and do it accurately. If you have to join the points up, then do this with a smooth curve.

- Make sure you are sure of what you are drawing. If you are asked to draw a histogram, then do not draw a polygon or cumulative frequency curve.

Read the questions thoroughly

- Do not try to save time by just skimming the questions. Read the questions carefully and highlight the important words. This should help you to remember to write down the answer that was asked in the question!

- It may seem obvious, but remember to answer the question that is asked. There will be no trick questions.

    For example:
    - If you are asked to write an equation in the form $ax + by = d$, where a, b and d are integers, then do not leave fractions or decimals in your answer.
    - If you are asked to write an answer to the nearest degree, then do not give the answer with decimal points in it even if this is correct to three significant figures.
    - If you are asked to give specific units, make sure you give these and that they are correct, for example m^2, cm, and so on.

- Read through the whole question more than once. There is often quite a lot of information provided in the equations and on the graph that you do not have to hunt for. Key words are sometimes written in bold, so that there is less chance of misinterpreting the question.

- Rely on your own thinking and knowledge rather than trying to remember how a similar question was solved in a sample exam. Sample exams are brilliant for helping you to prepare, but they will not be the same as the questions on the exam paper.

And finally

- Always check that your results are practicable, because examiners will not set questions with ludicrous answers. If your answer does not seem correct you should recheck your method and working. Have you ever seen a person 20 m tall and do you really think that London and New York are 20 cm apart?

5. Number and algebra

Number sets

You should be able to:

- understand the relationship between the sets: $\mathbb{N}$ (natural numbers), $\mathbb{Z}$ (integer numbers), $\mathbb{Q}$ (rational numbers) and $\mathbb{R}$ (real numbers)
- identify a given number as an element of one or more than one number set
- recognize the notation used for each of the number sets.

You should know:

- zero is a natural number.

Example

Mark with an ✕ the cell when the number is an element of the number set.

	$\mathbb{N}$	$\mathbb{Z}$	$\mathbb{Q}$	$\mathbb{R}$
3	✕	✕	✕	✕
−2		✕	✕	✕
$\frac{2}{3} = 0.\dot{6}$			✕	✕
$\sqrt{7}$				✕

Be prepared

- The decimal equivalent of a rational number will either have a finite number of decimal places or will repeat (periodic numbers).
- If a number is **not rational** then it is **irrational** and its decimal equivalent does not repeat or terminate.

For more information on number sets, see chapter 6.

Approximation

You should be able to:

- approximate a number to a given number of **decimal places** or **significant figures**.

Example

Use your GDC to calculate $\dfrac{2.45^5}{\sqrt{8}+1}$. Give your answer correct to:

(a) three significant figures;

When entering the expression in the calculator, bear in mind the use of brackets for the denominator.

Texas Instruments

```
2.45^5/(√(8)+1)
         23.05738426
```

Casio

MENU – RUN

```
2.45^5÷(√8+1)
         23.05738426
```

The calculator display shows 23.05738426. Therefore 23.1 is correct to three significant figures.

(b) three decimal places.

23.057 is correct to three decimal places.

Be prepared

- The decimal places are counted from the first figure after the decimal point, whereas the first significant figure is the first non-zero figure in the number.
- In intermediate steps, keep more than the accuracy required in the question, otherwise you will not reach the correct answer.

9

5. Number and algebra

Estimation

You should be able to:
- estimate a quantity
- use your estimations to check if your answer makes sense.

Example

Estimate the volume of a cube with a side length of 10.12 cm.

A possible estimate of the volume is $10^3 = 1000\,cm^3$

Percentage error

You should be able to:
- calculate percentage errors of measurements and estimations.

You should know:
- percentage error $= \dfrac{v_A - v_E}{v_E} \times 100\%$, where v_E represents the "exact value" and v_A represents the "approximated value" or "estimated value".

Example

The area of a room is $29.5\,m^2$. Antonio estimates the area as $30\,m^2$. Find the percentage error made by Antonio in his estimation.

Percentage error $= \dfrac{(30 - 29.5)}{29.5} \times 100\% = 1.69\%$

Be prepared
- Percentage errors do not have units.

Standard form (scientific notation)

You should be able to:
- write a number in the form $a \times 10^k$, where $1 \leq a < 10$ and $k \in \mathbb{Z}$.

Example

The number of bacteria in a culture is 234 000. Write this number in the form $a \times 10^k$, where $1 \leq a < 10$ and $k \in \mathbb{Z}$.

$234\,000 = 2.34 \times 10^5$

The International System of Units (SI)

You should be able to:
- identify the SI units and their derived units
- write down the correct combination of SI units when answering a question
- convert measurements from one unit to another (for example, km to cm, g to kg).

You should know:
- some of the units used in mathematical studies questions are metre (m), kilogram (kg), second (s), square metre (m²), metre per second (m/s or $m\,s^{-1}$).

Example

Paula runs 15 km in 2.5 h. Find Paula's average speed in $m\,s^{-1}$. Give your answer correct to two significant figures.

Average speed $= \dfrac{distance}{time} = \dfrac{15\,km}{2.5\,h} = \dfrac{15\,000\,m}{9000\,s} = 1.7\,m\,s^{-1}$

This is a sensible answer, whereas an average speed of $50\,m\,s^{-1}$ is not.

Be prepared
- Always check that your answers make sense. The distance from Earth to Pluto cannot be 10^3 km. The weight of a person cannot be 2300 kg.

Arithmetic and geometric sequences

You should be able to:
- use the formula for the nth term and for the sum of the first n terms
- solve real-world problems.

You should know:
- u_n represents the nth term (u_1 is the first term, u_2 is the second term, and so on)
- d is the **common difference** (the **difference** between a term and its previous term in an arithmetic sequence): $d = u_{n+1} - u_n$
- r is the **common ratio** (the **division** between a term and its previous term in a geometric sequence): $r = \dfrac{u_{n+1}}{u_n}$
- S_n is the sum of the first n terms: $S_n = u_1 + u_2 + \ldots + u_n$
- the nth term of an arithmetic sequence is $u_n = u_1 + (n-1)d$
- the sum of n terms of an arithmetic sequence is
$S_n = \dfrac{n}{2}(u_1 + u_n)$ or $S_n = \dfrac{n}{2}[2u_1 + (n-1)d]$
- the nth term of a geometric sequence is $u_n = u_1 r^{n-1}$
- the sum of n terms of a geometric sequence is
$S_n = \dfrac{u_1(1-r^n)}{1-r} = \dfrac{u_1(r^n-1)}{r-1}, r \neq 1$.

Example

The first term and seventh term of an arithmetic sequence are 3 and 24, respectively.

(a) Find the common difference.

Using the formula for the nth term $u_7 = u_1 + 6d$ then $24 = 3 + 6d$ and $d = 3.5$

(b) Find the 13th term.

$u_{13} = 3 + 12 \times 3.5$ so $u_{13} = 45$

(c) Find the sum of the first 13 terms.

Using the formula for the sum of n terms $S_{13} = \dfrac{13}{2}(2 \times 3 + 12 \times 3.5)$ therefore $S_{13} = 312$

Be prepared

- In an arithmetic sequence, each term is found by **adding** to the previous term a constant number (common difference). Arithmetic sequences are linked to simple interest within chapter 11.

- In a geometric sequence, each term is found by **multiplying** the previous term by a constant number (common ratio). Geometric sequences are linked to compound interest in chapter 11.

- Using the **formulae** instead of finding the answers by adding or multiplying will **save** you **time**.

Solutions of a pair of simultaneous equations

You should be able to:
- solve a pair of simultaneous equations in two variables by hand (presumed knowledge) or by use of the GDC.

You should know:
- a pair of linear simultaneous equations in two variables can be written as
  $ax + by = c$, $dx + ey = f$,
  where a, b, c, d, e, f are constants, and x and y are the two unknowns
- each of the two equations written above is the equation of a straight line, and the intersection of the lines is the solution of the pair of equations.

Example
Find the intersection between the lines $y = 2x + 3$ and $4x - y - 1 = 0$.

There are two methods. In both cases the GDC may be used.

- The algebraic method

  Rearrange to get

  $2x - y = -3$

  $4x - y = 1$

  and use the simultaneous equations facility.

Texas Instruments
APPS

Casio
MENU – EQUA

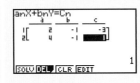

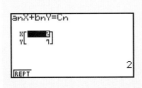

- The graphical method

  Rearrange the equations to make y the subject of the formulae

  $y = 2x + 3$

  $y = 4x - 1$

  and graph both lines to find the intersection point.

Texas Instruments

Graph lines in standard window.

CALC – intersect – place cursor near intersection point and press ENTER three times

Casio

Graph lines in standard window.

G-Solve – ISCT

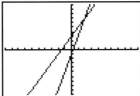

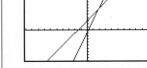

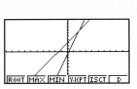

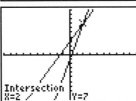

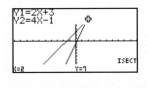

Solution is $x = 2$, $y = 7$.

Be prepared
- Check that the solution found satisfies the two equations given in the problem. If it does not, then you have made an error, either when rearranging the equations or when using your GDC.

Solution of quadratic equations

You should be able to:

- recognize a quadratic equation, which will always have an x^2 term in it
- solve a quadratic equation by factorizing.

You should know:

- any quadratic equation can be rearranged to the form $ax^2 + bx + c = 0$, where a, b, c are constants, $a \neq 0$, and x is the unknown
- if the product of two numbers is zero, then either one or both of the two numbers is zero
- the solutions to the equation $ax^2 + bx + c = 0$ are the roots of the quadratic function $f(x) = ax^2 + bx + c$
- the solution of a quadratic equation $ax^2 + bx + c = 0$ is
$$x = \frac{-b \pm \sqrt{b^2 - 4ac}}{2a}, a \neq 0$$
- the unknown of a quadratic equation may be named x, t, s, or any other letter.

Example

(a) Factorize the expression $t^2 + 2t - 8$.

The −8 can be written as the product of 1 and −8, or −1 and 8, or 2 and −4, or −2 and 4. From these pairs, −2 and 4 is the only pair that sums to 2. Therefore, the quadratic expression can be written as

$(t + 4)(t - 2)$

(b) Hence or otherwise, solve the equation $t^2 + 2t - 8 = 0$.

To solve the equation, you can either use part (a) or the GDC.

Using part (a), you know that the equation can be written as $(t + 4)(t - 2)$, so $t + 4 = 0$ or $t - 2 = 0$. Therefore, $t = -4$ or $t = 2$.

Using the GDC.

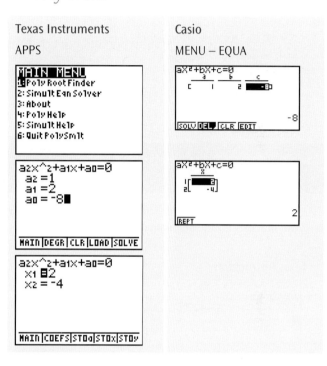

Be prepared

- If the question does not say "factorize", you can solve the equation using the GDC (this may save you time) or using the quadratic formula.

1. (a) Calculate $\dfrac{77.2 \times 3^3}{3.60 \times 2^2}$. [1 mark]

   (b) Express your answer to part (a) in the form $a \times 10^k$, where $1 \leq a < 10$ and $k \in \mathbb{Z}$. [2 marks]

   (c) Juan estimates the length of a carpet to be 12 metres and the width to be 8 metres. He then estimates the area of the carpet.

   (i) Write down his estimated area of the carpet. [1 mark]

   When the carpet is accurately measured it is found to have an area of 90 square metres.

   (ii) Calculate the percentage error made by Juan. [2 marks]

   [Taken from paper 1 November 2007]

How do I approach the question?

(a) You can use your GDC to calculate this expression. It is very important here to use brackets for the denominator.

Texas Instruments

(77.2*3^3)/(3.60*2²)
                144.75

Casio

(77.2×3^3)÷(3.60×2²)
                144.75

(b) This is linked to part (a), and you are asked to express your answer to (a) in standard form (scientific form). For this, you need to set your GDC in the appropriate mode. Remember to use mathematical notation (so, 1.45E+02 is not acceptable, as it is not mathematical notation). Also remember to give your answer correct to three significant figures. The answer is 1.45×10^2.

(c) (i) The key word here is "estimate". You just need to multiply two values to find the estimated area of the carpet.

   (ii) The key words here are "percentage error". You need to go to the information booklet and use the percentage error formula, with both 90, the accurate area (v_E), and 96, the estimated area (v_A).

What are the key areas from the syllabus?

- Arithmetic calculations through the use of the GDC
- Scientific notation
- Estimation
- Percentage errors

This answer achieved 2/6

It is clear that this student did not use brackets for the bottom of the expression.

Although part (a) is incorrect, there is correct follow-through from the student's answer, so the student is awarded full marks.

Although the student arrived at the correct answer, the student did not write m². So the UP was applied and no marks were awarded. This is a good example to highlight a clear lack of units costing the student 1 mark.

There is no working shown and the answer is incorrect. Therefore, 2 marks are lost.

Working:

Answers:

(a) *2316*   A0

(b) 2.316×10^3 C2ft

(c) (i) *96*   A0 (UP)

    (ii) *15%*   M0 A0

This answer achieved 3/6

The correct answer is found by opening a pair of brackets before entering the bottom part in the GDC.

The value for *k* is correct but the value *a* = 1.4 written in the expression $a \times 10^k$ is not given to three significant figures, for which the student is penalized with an AP.

Correct area found by multiplying 12 and 8, and the student has added m².

There is no working out shown and incorrect answer written, so no marks gained.

Working:

144.75

1.4×10^2

$\dfrac{96}{90}$ M0

Answers:

(a) *144.75*   C1

(b) 1.4×10^2 A0 AP A1

(c) (i) $96 \, m^2$ C1

    (ii) *−6.67%*   A0

This answer achieved 6/6

Instead of doing just one calculation, this student found the value for the top, then the value for the bottom and then divided the two.

The answer was correctly rounded to three significant figures when writing down the answer in the answer space.

Although the working is incorrect (the substitution into the percentage error formula is wrong), the answer given is correct. As this is a paper 1, the rule that the examiners follow is that full marks should be awarded to a correct answer independently of the working shown.

Working:

1. (a) $77.2 \times 3^3 = 2084.4$

   $3.60 \times 2^2 = 14.4$

   $\dfrac{2084.4}{14.4} = 144.75$

   (b) 1.45×10^2

   (c) [rectangle: 12 m × 8 m]

   (i) $8 \times 12 = 96 \, m^2$

   (ii) $\% = \dfrac{90 - 96}{90}$

   $= 6.67$

Answers:

(a) $145 \; 3sf$ — C1

(b) 1.45×10^2 — C2

(c) (i) $96.0 \, m^2$ — A1

   (ii) $6.67\% \; 3sf$ — C2

Examiner report

In part (a) a common mistake was to enter the numbers in the GDC without using brackets and arrive at the wrong answer. Also be aware of losing marks for an accuracy penalty.

Students can often work out the estimated area of the carpet in (c)(i). However, a common mistake is forgetting to add the unit (m^2), and they lose 1 mark. Another problem is that students often incorrectly use the percentage error formula in (c)(ii) and substitute the incorrect values. You need to be sure that you are familiar with the meaning of v_A and v_E in this formula. Also after having found the correct value, students often forgot to give the answer correct to three significant figures, so they lose an accuracy mark if they had not lost one previously in the question.

To help with this type of question, students should clearly show their working out for each single part question.

10. (a) Solve the following equation for x

$$3(2x+1)-2(3-x)=13.$$

[2 marks]

(b) Factorize the function x^2+2x-3.

[2 marks]

(c) Find the **positive** solution of the equation

$$x^2+2x-6=0.$$

[2 marks]

[Taken from paper 1 November 2007]

How do I approach the question?

(a) This requires a linear equation in one variable to be solved. You need to expand both terms. Be very careful when expanding the second term with the minus sign in front of the 2.

(b) You need to find two integers that multiplied together give -3 and that sum to 2. These integers are 3 and -1. So the factorized expression for x^2+2x-3 is $(x+3)(x-1)$.

(c) Again, you are required to solve an equation. But this one is a quadratic equation, whereas in (b) you have been asked to factorize a similar one. From the wording of the question, it is clear that the equation has two solutions, a positive one and a negative one. The answer will be the **positive** solution. To help the student, this is highlighted in bold. You can solve the equation by use of the GDC or by use of the formula. As this question is towards the end of the paper, it would save you valuable time if you use the GDC.

Texas Instruments

APPS

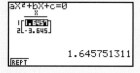

Casio

MENU – EQUA

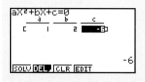

Which GDC functions should I use?

(a) You can use solver to solve the equation.

Texas Instruments

MATH – SOLVER

Casio

MENU – EQUA – SOLVER

What presumed knowledge should I have?

You should know how to expand simple algebraic expressions and how to solve linear equations in one variable.

This answer achieved 2/6

This student makes a mistake when expanding the brackets in the second term of the equation. The student considers the minus in front of the 2 just when multiplying by 3 but not when multiplying by x, which also has a minus in the front.

Not only is the answer correct, but the good thing here is that the student checked that, when expanding the brackets, the original expression was obtained.

Showing working in the answer space is not a good thing. There is a special area designated for the working out! Where does the x = 2 come from? It seems that the student chose it at random. Why is the student happy with this value for x if it does not verify the equation? In fact, if this were the correct solution, the right-hand side of the equation should be 0 and not 2.

Working:

$3(2x + 1) - 2(3 - x) = 13$

$6x + 3 - 6 - 2x = 13$ M0

$6x - 2x + 3 - 6$

$4x + (-3) = 13$ $x^2 + 2x - 3$

$4x = 13 + 3$ $(x - 1)(x + 3)$

$4x = 16$ $x^2 + 3x - x - 3$

$x = \frac{16}{4}$ $x^2 + 2x - 3$

$x = 4$

Answers:

(a) $x = 4$ A0

(b) $(x - 1)(x + 3)$ A2

(c) $2^2 + 2 \times 2 - 6 = 2$

$x = 2$ M0 A0

This answer achieved 4/6

This part of the answer highlights that the student has a good understanding of algebra. Brackets are well expanded and there is good simplification of the expression to finally find the solution.

The quadratic expression is written as a product of the correct two linear expressions.

Here it seems that the student misunderstood the question. Instead of solving the equation, the student has differentiated and found the values of x for which the gradient of the function is positive.

Working:

(a) $3(2x + 1) - 2(3 - x) = 13$

$6x + 3 - 6 + 2x = 13$

$8x - 3 = 13$

$8x = 16$

$x = 2$

min. at $(-1, -7)$

(b) $f(x) = x^2 + 2x - 3$

$= (x - 1)(x + 3)$

(c) $f(x) = x^2 + 2x - 6$

$f'(x) = 2x + 2$

$0 = 2x + 2$

$2 = 2x$

$x = -1$

Answers:

(a) $x = 2$ C2

(b) $f(x) = (x - 1)(x + 3)$ C2

(c) *positive gradient when* $x > -1$ M0 A0

This answer achieved 6/6

Working:

(a) $3(2x + 1) - 2(3 - x) = 13$

$6x + 3 - 6 + 2x = 13$

$8x - 3 = 13$

$8x = 16$

$x = 2$

(b) $x^2 + 2x - 3$

$x^2 + 3x - 1x - 3$

$= x(x + 3) - 1(x + 3)$

$= (x + 3)(x - 1)$

(c) $a = 1, b = 2, c = -6$

$$x = \frac{-2 + \sqrt{2^2 - 4(1 \times -6)}}{2 \times 1}$$

$$= \frac{-2 + \sqrt{28}}{2}$$

$$= 1.645751$$

Answers:

(a) $x = 2$ — C2

(b) $(x + 3)(x - 1)$ — C2

(c) $x = 1.65$ (3SF) — C2

- Excellent manipulation of the algebra. The student shows clear working step by step.
- Good use of the quadratic formula. The coefficients a, b and c from the quadratic formula are clearly identified before starting to use it.
- The student managed to factorize the quadratic expression by grouping terms.
- Student correctly rounded to three significant figures.

Examiner report

In this question, you had to know how to solve linear and quadratic equations and also how to factorize quadratic equations. A common error is that students do not use the rule that minus times minus gives a plus. This leads to the linear equation not being solved correctly.

If you are unsure of the key word(s) within a question, take some time to refresh your memory. For example, not understanding "factorize" meant that students gave the answer $x(x + 2) - 3$. But this expression is not factorized, as it is not written as a product.

The last part of the question should take you very little time if you find the solution by use of the GDC. However, take your time and read the question correctly. Some students misread the question and gave the positive solution to the equation given in (b).

4. *[Maximum mark: 16]*

(i) The natural numbers: 1, 2, 3, 4, 5… form an arithmetic sequence.

(a) State the values of u_1 and d for this sequence. *[2 marks]*

(b) Use an appropriate formula to show that the sum of the natural numbers from 1 to n is given by $\frac{1}{2}n(n+1)$. *[2 marks]*

(c) Calculate the sum of the natural numbers from 1 to 200. *[2 marks]*

(ii) A geometric progression G_1 has 1 as its first term and 3 as its common ratio.

(a) The sum of the first n terms of G_1 is 29 524. Find n. *[3 marks]*

A second geometric progression G_2 has the form $1, \frac{1}{3}, \frac{1}{9}, \frac{1}{27}\ldots$

(b) State the common ratio for G_2. *[1 mark]*

(c) Calculate the sum of the first 10 terms of G_2. *[2 marks]*

(d) Explain why the sum of the first 1000 terms of G_2 will give the same answer as the sum of the first 10 terms, when corrected to three significant figures. *[1 mark]*

(e) Using your results from parts (a) to (c), or otherwise, calculate the sum of the first 10 terms of the sequence $2, 3\frac{1}{3}, 9\frac{1}{9}, 27\frac{1}{27}\ldots$
Give your answer **correct to one decimal place**. *[3 marks]*

[Taken from paper 2 May 2007]

How do I approach the question?

This question is made up of two parts, both concentrating on separate areas of the syllabus.

(i) You are told that the sequence is arithmetic.
  (a) You need to have a clear idea of the meaning of u_1 and d.
  (b) The key word is "sum", so use the formula for the sum of n terms of an arithmetic sequence, with the correct values (there are n terms from 1 to n and the common difference, d, is 1).
  (c) Again, the key word is "sum". Use the formula given in (b) to find the answer to (c). You just substitute n by 200.

(ii) You are introduced to two geometric sequences, G_1 and G_2, for which you are given different information. Identify the key words in each part question as "common ratio" and "sum" and then relate to their meaning and/or formula to find them.
  (a) You are given the sum and have to find the number of terms, n. Substitute into the sum formula to set the equation and go to find n. The equation may seem difficult but you can use your GDC to solve it.

What are the key areas from the syllabus?

- Sequences.

How does this relate to the information formula booklet?

You should be aware that in the information formula booklet there are six formulae related to sequences. You need to identify which formula to use. Once you have identified the formula, you need to understand how to substitute the values from the problem in these formulae.

Texas Instruments

MATH – SOLVER, enter equation then press ENTER – ALPHA ENTER (SOLVE)

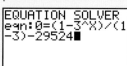

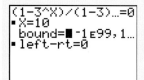

Casio

MENU – EQUATION – Solver – enter equation – EXE then press SOLVE

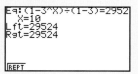

  (b) To find the common ratio, divide each term by its previous terms and you will see that it always gives $\frac{1}{3}$.
  (c) Use the formula for the sum again, but this time you know that $n = 10$ and need to find the sum.
  (d) This is a reasoning question, which means that you have to justify or explain why something happens.
  (e) You are told to look at the results found in (a) to (c) to find the answer as the new sequence is a "combination" of G_1 and G_2. Look at the sequence and you will notice that, to find the sum of its first 10 terms, you can just add the sum of the first terms of G_1 and the sum of the first terms of G_2.

This answer achieved 4/16

The student correctly identified the values of u_1 and of d.

There is a formula copied from the booklet, but you need to substitute the right formula with values from the problem to have some credit. In this case the student did not realize that the key word was "sum" and started to work with the formula for the nth term.

The formula for the sum is used and the correct substitution made, so full marks are awarded.

In the question it is clearly stated that the sequence is **geometric** and it is said that the **sum** is 29 524. However, the student uses the formula for the nth term of an arithmetic sequence.

Working:

4 (i) (a) $u_1 = 1 \; d = +1$ A1 A1

     (b) $u_n = u_1 + (n-1)d$ M0 A0

     (c) $S_n = \frac{n}{2}(2u_1 + (n-1)d)$ M1

        $S_n = \frac{200}{2}(2(1) + (200-1)1)$

        $S_n = 100(2 + 199)$

        $S_n = 100(201)$

        $S_n = 20\,100$ A1

(ii) (a) $u_n = u_1 + (n-1)d$

       $29\,524 = 1 + (n-1)3$ M0

       $29\,524 = n - 1 - 3$ A0

       $29\,524 = n - 3$

       $n = 29{,}527$

This answer achieved 10/16

> The student gains full marks for the first part of the question. Clear working has been shown, with good use of the formula for the sum of the first n terms of an arithmetic sequence.

> Good use of the formula for the sum of the first n terms of a geometric sequence.

> The student makes an algebraic mistake when solving the equation $3^n = 59\,049$, and does not find the correct value for n. Is the value 38.9 possible for n? What kind of number is n?

> The student has found the incorrect common ratio. The common ratio is the division between any term and its previous term (u_2/u_1, u_3/u_2, and so on).

> Although the answer is incorrect, it does follow through from the student's previous answer, and so is awarded full marks.

> The main mistake here was to take the new sequence as geometric, whereas it is not. Does the found value make sense? You need to sum the terms 2, $3\frac{1}{3}$, ..., so the value 3.2 is not possible!

Working:

4 (i) (a) $u_1 = 1$ A1
         $d = 1$ A1

(b) $S_n = \dfrac{n}{2}(2u_1 + (n-1)d)$

     $= \dfrac{n}{2}(2 \times 1 + (n-1) \times 1)$ M1

     $= \dfrac{n}{2}(2 + (n-1))$ A1

(c) $S_n = \dfrac{200}{2}(2 \times 1 + (200-1)1)$ M1

     $= 100(2 + 199)$

     $= 100 \times 201$

     $= 20\,100$ A1

(ii) (a) $u_1 = 1$
        $r = 3$

        $29\,524 = \dfrac{1(3^n - 1)}{3 - 1}$ M1 A1

        $29\,524 = \dfrac{3^n - 1}{2}$

        $59\,048 = 3^n - 1$

        $59\,049 = 3^n$

        $\sqrt[3]{59\,049} = 38.9$

        $n = 38.9$ A0

(b) $\dfrac{1}{3^3} = r$ A0

(c) $S_n = \dfrac{1\left(\dfrac{1}{3^3}\right)^{10} - 1}{\dfrac{1}{3^3} - 1}$ M1

     $= \dfrac{-1}{-0.9629}$

     $= 1.09$ A1ft

(d)
(e) $S_n = \dfrac{2\left(3\dfrac{1}{3^3}\right)^{10} - 1}{3\dfrac{1}{3^3} - 1}$

     $= \dfrac{2(9.856 - 1)}{-0.6296}$ M0

     $= \dfrac{-1.9}{-0.6296}$

     $= 3.17 = 3.2$ A0

This answer achieved 14/16

The student uses the expression given in the question and substitutes *n* by 1, which is just a particular value for *n*. What you have to do here is to "show that" the formula works for any value of *n* and not just for one particular value of *n*.

Good use of the formula for the sum and excellent working to find the value of *n*.

The student keeps more than three significant figures in the intermediate steps, and in the final step the answer is given correct to three significant figures.

It is clear that the student has an idea, though it is not well explained. But the student achieves one mark for reasoning. The sum of the first 1000 terms is calculated and compared with the sum of the first 10 terms. Also, the idea of the terms getting smaller and smaller and making no difference in the total sum is mentioned.

Good thinking! The student constructed the 10 terms of the sequence and added all the terms up.

Working:

4 (i) (a) $u_1 = 1$ A2
$d = 1$

(b) $\frac{1}{2}n(n+1)$ M0

$n = 1$ A0

$\frac{1}{2}(1)((1)+1) = 1$

(c) $S_{200} = \frac{200}{2}(2(1) + (200-1)1)$ M1

$S_{200} = 20\,100$ A1

(ii) (a) $29\,524 = \frac{1(3^n - 1)}{3-1}$ M1

$59\,048 = 1(3^n - 1)$
$59\,048 = 3^n - 1$
$59\,047 = 3^n$
$n = 10$ A1 A1

(b) $r = \frac{1}{3}$ A1

(c) $S_{10} = \frac{1\left(\frac{1}{3}^{10} - 1\right)}{\frac{1}{3} - 1}$ $S_{10} = 1.50$ M1 A1

$= \frac{-0.99998}{-\frac{2}{3}}$

(d) $S_{1000} = \frac{1\left(\frac{1}{3}^{1000} - 1\right)}{\frac{1}{3} - 1}$

$S_{1000} = 1.50$

Because the terms in G_2 are decreasing by $\frac{1}{3}$ and therefore do not make much of a difference in the overall sum.   R1

(e) $2, 3\frac{1}{3}, 9\frac{1}{9}, 27\frac{1}{27}, 81\frac{1}{81}, 243\frac{1}{243}, 729\frac{1}{729}, 2187\frac{1}{2187}, 6561\frac{1}{6561}, 19683\frac{1}{19683}$ M1

$S_{10} = 29525.5$ A2

Examiner report

For part (i) as long as you know how to identify the values u_1 and d there should be no problem. In the "show that" part, students often find it difficult to understand that they have to show the formula worked for **all** values of *n* rather than proving that the formula given worked for one particular value of *n*.

Within part (ii) it was important to read the question carefully. The geometric sequences appeared here and the formulae needed were different from those needed in part (i). If you were aware of the information given in each part question, then parts (a) to (c) should not give you too much trouble. The majority of the students found it difficult to give a good explanation in (d). What they had to do was to compare $\left(\frac{1}{3}\right)^{10}$ and $\left(\frac{1}{3}\right)^{1000}$ and observe that both values are 0 when corrected to three significant figures, and so there is no difference in S_{10} and S_{1000}. The last part was the most difficult one in the question. Not many students noticed that the *n*th term of this sequence was the addition of the *n*th terms of each of G_1 and G_2. Those students who worked out each of the 10 terms of this new sequence and then added them up were successful.

6. Sets, logic and probability

Basic concepts of set theory

You should be able to:

- understand and give examples of the following concepts: set, subset, empty set, universal set, complement of a set, intersection and union
- list or describe the elements of a given set
- identify the elements that belong to the subset of a given set
- decide whether a statement about a given set or sets is correct or incorrect
- use the common set notation
- use set notation to describe number sets (for more information on number sets see chapter 5)
- translate a verbal statement involving given sets into a statement written in set notation and vice versa.

You should know:

- $A = \{a, b, c, d, ..., n\}$ means that the elements of A are $a, b, c, d, ...$ and n
- $A = \{x \in \mathbb{N} \mid a < x < b\}$ means that the elements of A are all natural numbers between a and b
- $d \in A$ means that d is an element of set A
- $y \notin A$ denotes that y is not an element of A
- a group of elements of $A = \{a, b, c, d, ..., n\}$, for example $B = \{a, c\}$, is a proper subset of A, and we write $B \subset A$
- the symbol $\varnothing$ denotes the empty set
- the universal set is denoted by U
- A' denotes the complement of a set A to a given universal set U
- $A \cup B$ denotes the union of sets A and B
- $A \cap B$ denotes the intersection of sets A and B.

Example

Let U be the set of all positive integers from 1 to 11 inclusive, and M and N be the subsets of U where: M is the set of all the positive integers that are not multiples of 5 and 3, and N is the set of the even numbers.

(a) List all the elements of set M.

$M = \{1, 2, 4, 7, 8, 11\}$

(b) List the set notation to describe $M \cap N'$.

Since $N = \{2, 4, 6, 8, 10\}$ and $N' = \{1, 3, 5, 7, 9, 11\}$, then $M \cap N' = \{1, 7, 11\}$

Be prepared

- You may find it easier to solve such problems by visualizing the sets.
- The order of the elements of a set is not important. But it might help you when you work with subsets of $\mathbb{N}$ or $\mathbb{Z}$ if you order their elements.
- When you define a complement of a given set, keep in mind which your universal set is.

Venn diagrams

You should be able to:

- use Venn diagrams to represent a set within a given universal set, subsets of a set, intersection of sets, union of sets and disjoint sets
- use Venn diagrams to identify the elements of a complement of a given set, subset, union, or intersection of given sets
- shade regions on a Venn diagram that represent various subsets
- use set notation to describe a set that is represented by a shaded region on a Venn diagram.

You should know:

- $n(A)$ denotes the number of elements of set A.

Example

The following is the result of a brief survey of a group of year 8 students that shows what their favourite after-school activity is. These are the responses:

21 like watching films
12 like jogging only
5 like watching films and jogging
4 do not like doing any of the activities listed above.

(a) Draw a Venn diagram and enter the above information in the appropriate places.

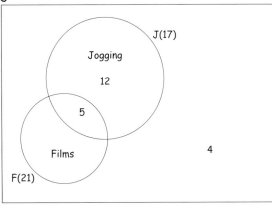

(b) Find how many students like to watch films only.

The number of students who like to watch films only is 21 − 5 = 16

(c) Find how many students responded in the survey.

The total number is 21 + 12 + 4 = 37

Be prepared

- Begin working on your Venn diagram by identifying, drawing and labelling the universal set.
- The universal set is represented by a rectangle and labelled U, and the subsets are represented by circles.

Basic concepts of symbolic logic: propositions

You should be able to:

- recognize a proposition
- distinguish between simple and compound propositions
- recognize and use logical connectives to form a compound proposition
- translate from verbal representation of a proposition to a symbolic representation and vice versa.

You should know:

- a simple proposition is a meaningful declarative sentence
- often simple propositions are denoted by p, q or r
- compound propositions are formed from simple ones with the use of the logical connectives
- the connectives that are listed in the following table.

Name	Symbol
"not" negation	$\neg$
"and" conjunction	$\wedge$
"or" (inclusive) disjunction	$\vee$
"or ... but not both" exclusive disjunction	$\underline{\vee}$
"if ... then" implication	$\Rightarrow$
"if and only if" equivalence	$\Leftrightarrow$

Example

(a) Simple propositions

Dragons exist.

Jonathan Swift is the author of Gulliver's Travels.

(b) Examples of sentences that are not propositions

What a beautiful flower!

Where does this road lead?

(c) Compound propositions

If it rains then I ride a bus to school.

p : It rains.

q : I ride a bus to school.

Here p and q are simple propositions, if ... then is the connective, and the entire compound proposition is $p \Rightarrow q$.

Be prepared

- You could identify the simple propositions in a compound proposition by first identifying and perhaps circling the logical connectives.
- To translate a verbal expression into a symbolic one, identify all simple propositions, name them (for example, p, q), and use those names for the simple propositions and the symbols for the connectives.

Compound propositions, truth tables, tautology and contradiction

You should be able to:

- identify the following compound propositions: negation, conjunction, disjunction, exclusive disjunction (often known as "or but not both"), and implication
- write down examples of those compound propositions in symbolic or verbal form
- write down the truth tables for those compound propositions or other unfamiliar ones
- draw a Venn diagram and identify the truth set for a given proposition
- determine that a statement is a contradiction by evaluating its truth table
- determine that a statement is a tautology by evaluating its truth table.

You should know:

- $(\neg p)$ is commonly used for the logical negation of proposition p, $(p \wedge q)$ for conjunction, $(p \vee q)$ for disjunction, $(p \veebar q)$ for exclusive disjunction ("or but not both"), and $(p \Rightarrow q)$ for implication.

Example

Given the statements:
p : It rains.
q : I am carrying my umbrella.

(a) Write down, in words, the meaning of $\neg p \Rightarrow \neg q$.

If it does not rain, then I am not carrying my umbrella.

(b) Complete the truth table.

p	q	$p \wedge q$	$\neg p$	$\neg q$	$\neg p \vee \neg q$
T	T	T	F	F	F
T	F	F	F	T	T
F	T	F	T	F	T
F	F	F	T	T	T

(c) Show that $(p \wedge q) \vee (\neg p \vee \neg q)$ is a tautology.

Construct a truth table for the proposition $(p \wedge q) \vee (\neg p \vee \neg q)$.

$p \wedge q$	$\neg p \vee \neg q$	$(p \wedge q) \vee (\neg p \vee \neg q)$
T	F	T
F	T	T
F	T	T
F	T	T

Since the last column shows that it can take only T values, this proposition is a tautology.

Be prepared

- To verify that a proposition is a tautology, you could construct the truth table.
- If, in a truth table, the column for the values of a proposition contain only F values, then this proposition is a logical contradiction.
- When expressing a proposition symbolically, use brackets for each compound proposition that is contained in another.

Related statements, logical equivalence, converse, inverse and contrapositive

You should be able to:
- identify whether two propositions are equivalent by comparing their truth values
- write down the converse, inverse and contrapositive of a given implication
- give simple examples of a converse, inverse and contrapositive
- show that the converse and inverse are logically equivalent
- show that an implication and its contrapositive are logically equivalent.

You should know:
- logical equivalence of statements p and q is denoted $p \Leftrightarrow q$
- implication is denoted by $p \Rightarrow q$, which is read as "if p then q"
- the inverse of $p \Rightarrow q$ is denoted by $\neg p \Rightarrow \neg q$, which is read as "if not p then not q"
- the converse of $p \Rightarrow q$ is $q \Rightarrow p$, which is read as "if q then p"
- the contrapositive of $p \Rightarrow q$ is $\neg q \Rightarrow \neg p$, which is read as "if not q then not p".

Example
Consider the statements:
p : There is no traffic.
q : I will catch the train.

(a) Write down the inverse of the statement $p \Rightarrow q$.

If there is traffic, then I will not catch the train.

(b) Show that the implication $p \Rightarrow q$ and its contrapositive are logically equivalent.

Construct the truth table for $p \Rightarrow q$ and the contrapositive $\neg q \Rightarrow \neg p$.

p	q	$p \Rightarrow q$	$\neg q$	$\neg p$	$\neg q \Rightarrow \neg p$
T	T	T	F	F	T
T	F	F	T	F	F
F	T	T	F	T	T
F	F	T	T	T	T

Comparing the columns of the two propositions (columns 3 and 6) highlights that they are the same. Therefore the two propositions in question are logically equivalent.

Be prepared
- Often people think that the inverse $\neg p \Rightarrow \neg q$ is logically equivalent to $p \Rightarrow q$. This is not true! Use truth tables and check for yourself.

Events, sample space and probability

You should be able to:
- identify the sample space of an experiment, that is, the set of all possible outcomes
- calculate the probability of an event occurring
- calculate the probability of a complementary event.

You should know:
- $P(A)$ denotes the probability that a particular event A will occur by chance
- A and A' denote two complementary events, U denotes the sample space
- probability of an event A, $P(A) = \dfrac{n(A)}{n(U)}$
- probability of a complementary event A', $P(A') = 1 - P(A)$

Example

(a) What is the probability of choosing a girl in a single random selection from a class of 24 students with 15 girls in?

It is: $P(G) = \dfrac{15}{24}$.

(b) What is the probability of choosing a boy from the same class?

It is: $P(B) = 1 - \dfrac{15}{24} = \dfrac{9}{24}$, since G and B are complementary events.

Be prepared
- The sum of the probability of all possible outcomes is one.
- If something is certainly going to happen, then the probability is 1.

Venn and tree diagrams in probability

You should be able to:

- use Venn diagrams, when appropriate, to calculate probabilities of events
- represent alternative events and their probabilities using a tree diagram when feasible
- use a tree diagram to represent situations when answering probability questions.

Example

A jar is filled with 20 marbles, of which 11 are red and 9 are blue. Walter removed two marbles at random.

(a) Draw a tree diagram to represent the experiment.

The second marble is taken after the first one has been removed and not replaced. Thus this is a problem "without replacement."

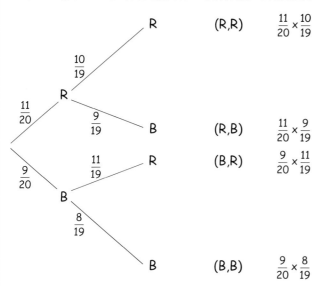

(b) Find the probability that:
(i) both marbles are blue

The probability is

$P(B, B) = \frac{9}{20} \times \frac{8}{19} = \frac{18}{95}$

The events of choosing a blue marble and then choosing a blue marble again are represented by the lowest branch in the tree diagram.

(ii) one marble is blue and one marble is red.

The probability is:

P(one marble is blue and one marble is red) =
$P(R, B) + P(B, R)$

since the order of the blue and red marbles does not matter here. So

P(one blue and one red) $= \frac{11}{20} \times \frac{9}{19} + \frac{9}{20} \times \frac{11}{19} = \frac{99}{190}$

Be prepared

- In problems "without replacement", the sample space for the second event changes.
- In problems "with replacement", the sample space for the second event does not change.
- Tree diagrams are usually used in probability problems with alternative events.
- The probability of the outcome of a series of events in a branch is calculated by multiplying the probabilities of the events along the path of the branch.
- The sum of the probabilities of the outcomes in each of the branches (the sum of the numbers in the "Probability" column) in a tree diagram is one.

Laws of probability

You should be able to:

- understand and give examples of the following concepts: mutually exclusive events, independent events, combined events, and an event that occurs given that another event has already occurred
- use the formulae to calculate the probability of combined events, mutually exclusive events, or independent events, and conditional probability.

You should know:

- probability of combined events A and B

$$P(A \cup B) = P(A) + P(B) - P(A \cap B)$$

- probability of combined events A and B if A and B are mutually exclusive

$$P(A \cup B) = P(A) + P(B)$$

- probability of independent events

$$P(A \cap B) = P(A) \times P(B)$$

- conditional probability

$$P(A \mid B) = \frac{P(A \cap B)}{P(B)}$$

Example

For a class of 26 students, the options for extracurricular activities are an art class and a photography class. 15 students attend the art class, 10 students attend the photography class, and 4 students attend neither. Find

(a) the probability that a student chosen at random attends both the art and the photography class

Draw a Venn diagram.

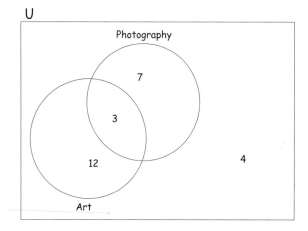

$P(\text{a student attends both classes}) = \frac{3}{26}$

(b) the probability that a randomly chosen student attends the photography class knowing that the student attends the art class.

$P(\text{photography class} \mid \text{art class})$

$$= \frac{P(\text{student attends both classes})}{P(\text{student attends art class})}$$

$$= \frac{3}{15}$$

Be prepared

- First try to visualize the information given in the problem (Venn diagram, tree diagram, table), and then identify the formulae that you might need to answer the questions.

4. The truth table below shows the truth-values for the proposition

$$(p \underline{\vee} q) \Rightarrow (\neg p \underline{\vee} \neg q)$$

p	q	$\neg p$	$\neg q$	$p \underline{\vee} q$	$\neg p \underline{\vee} \neg q$	$(p \underline{\vee} q) \Rightarrow (\neg p \underline{\vee} \neg q)$
T	T	F	F		F	
T	F	F		T	T	T
F	T	T	F	T	T	T
F	F	T	T	F		T

(a) Explain the distinction between the compound propositions, $p \underline{\vee} q$ and $p \vee q$.

(b) Fill in the four missing truth-values on the table.

(c) State whether the proposition $(p \underline{\vee} q) \Rightarrow (\neg p \underline{\vee} \neg q)$ is a tautology, a contradiction or neither.

[Taken from paper 1 May 2007]

How do I approach the question?

(a) You have to understand clearly the difference between disjunction and exclusive disjunction in order to explain how they differ.

(b) The table is to be completed step by step, using your knowledge of negation, exclusive disjunction and implication.

(c) You have to look at the last column of the table where the truth values of the given proposition are to be found. This will help you to determine which proposition it is.

What are the key areas from the syllabus?

- Compound propositions
- Exclusive disjunction
- Negation
- Implication
- Truth tables
- Tautology
- Contradiction

Please be aware that we have replicated the table within the working area to show each student's answer. The student was not expected to do this and had to mark up the table within the question.

6. Sets, logic and probability

This answer achieved 2/6

The explanation does not show a clear understanding of the distinction between disjunction and exclusive disjunction.

All values placed in the table are incorrect. A follow-through mark is awarded for the entry in the last column. This last value is incorrect, but it would have been correct if the student's entry in column 5 had been correct.

Although the conclusion drawn by the student is incorrect, it is correct given the truth values of the proposition in the last column of the student's table. As a result, the student is awarded a follow-through mark.

Working:

(a) $p \vee q$ means p 'or' q

$p \veebar q$ means p

(b)

p	q	$\neg p$	$\neg q$	$p \veebar q$	$\neg p \veebar \neg q$	$(p \veebar q) \Rightarrow (\neg p \veebar \neg q)$	
T	T	F	F	T	F	F	A0 A1ft
T	F	F	F	T	T	T	A0
F	T	T	F	T	T	T	
F	F	T	T	F	T	T	A0

(c) $p \veebar q \Rightarrow \neg p \veebar \neg q$

is neither a tautology (always true) or a contradiction (never true)

Answers:

(a) $p \vee q$ means p 'or' q

$p \veebar q$ means   A0

(c) Neither   A1 ft

This answer achieved 3/6

The explanation does not show any understanding of the difference between disjunction and exclusive disjunction. Moreover, there seems to be a misconception about what disjunction is.

Here 1 mark is given for the correct value in column 4, and another is a follow-through mark given for the value in column 7.

Although the conclusion drawn by the student is incorrect, it is correct given the truth values of the proposition in the last column of the student's table. As a result, the student is awarded a follow-through mark.

Working:

$t \to f - f \qquad f \vee f = f \qquad t \wedge t = t$

(b)

p	q	$\neg p$	$\neg q$	$p \veebar q$	$\neg p \veebar \neg q$	$(p \veebar q) \Rightarrow (\neg p \veebar \neg q)$	
T	T	F	F	T	F	F	
T	F	F	T	T	T	T	A2
F	T	T	F	T	T	T	
F	F	T	T	F	T	T	

Answers:

(a) $\veebar$ is and/or

while $\vee$ is and   A0

(c) Neither   A1ft

This answer achieved 6/6

The student has a correct and clear explanation. This highlights the understanding of the difference between the two propositions.

All four entries in the truth table are correct.

The student makes the correct conclusion that the given proposition is a tautology.

Working:

$\wedge$ = and

$\vee$ = or

$p \veebar q$ means only p or only q is true,

$p \vee q$ means either p or q or both are true

(b)

p	q	$\neg p$	$\neg q$	$p \veebar q$	$\neg p \veebar \neg q$	$(p \veebar q) \Rightarrow (\neg p \veebar \neg q)$
T	T	F	F	F	F	T
T	F	F	T	T	T	T
F	T	T	F	T	T	T
F	F	T	T	F	F	T

C4

Answers:

(a) $p \veebar q$ means only p or only q is true.   A1
$p \vee q$ means either p or q or both are true.

(c) Tautology   A1

Examiner report

A common error was that students did not show a thorough understanding of exclusive disjunction and of disjunction. Moreover, many students who seemed to be familiar with exclusive disjunction were not always able to articulate clearly the difference between disjunction and exclusive disjunction. Explanations were mostly given in phrases like "the second is p or q" and "the first is p or q but not both". When explanations are asked for, they ideally need to be given in complete sentences, and not just offered as single phrases. For example, exclusive disjunction ($p \veebar q$) is an operation that produces true values if and only if p or q, but not both of them, are true. Whereas disjunction ($p \vee q$) always produces true values except for the case in which both p and q are false. As for (c), a correct conclusion was relatively easy to identify, although sometimes it was inferred on the basis of their incorrect entries in the truth table.

7. B and C are subsets of a universal set U such that

$$U = \{x : x \in \mathbb{Z}, 0 \leq x < 10\}, B = \{\text{prime numbers} < 10\}, C = \{x : x \in \mathbb{Z}, 1 < x \leq 6\}.$$

(a) List the members of sets

   (i) B

   (ii) $C \cap B$

   (iii) $B \cup C'$

Consider the propositions:

$p : x$ is a prime number less than 10.
$q : x$ is a positive integer between 1 and 7.

(b) Write down, in words, the contrapositive of the statement, "If x is a prime number less than 10, then x is a positive integer between 1 and 7."

[Taken from paper 1 May 2007]

How do I approach the question?

(a) (i) Make a list of the elements of U, B and C first.

   (ii) Once you have identified the elements of B and C, you will be able to find the elements of $C \cap B$.

   (iii) Before you make a list of the elements of $B \cup C'$, you need to define C' and list its elements. Then you will be able to identify the members of $B \cup C'$ and make a list of them.

(b) You have to know what a contrapositive is, and to write down the contrapositive of the statement that is given.

What are the key areas from the syllabus?

- Universal set
- Subset
- Complement of set
- Union
- Intersection of sets
- Contrapositive

What presumed knowledge should I have?

- You should know what a prime number is and what symbolic expressions such as $x > 1$ or $x \geq 0$ mean.

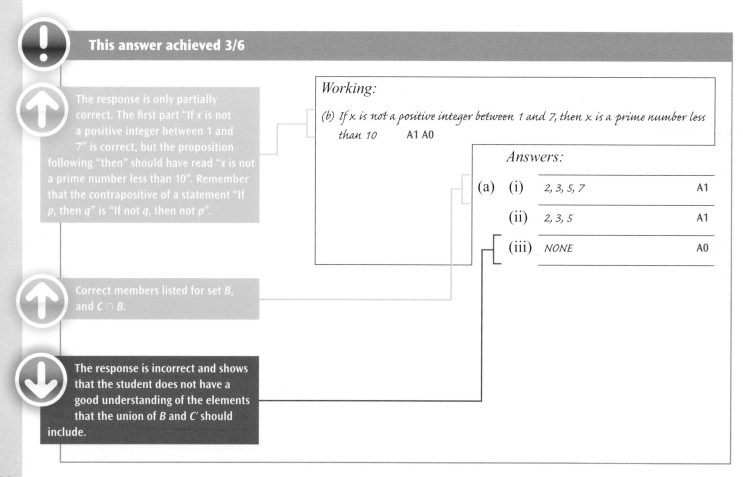

This answer achieved 6/6

The U is incorrect, but did not affect the rest of the student answer or any of the marks as it was not asked for.

The student first listed the elements of B, then the elements of C, which is a logical step to take before one is able to identify the elements of $C \cap B$.

Full marks awarded for correctly identifying the elements of C' and the correct list of the elements of $B \cup C'$.

The student first expressed the contrapositive symbolically ($\neg q \Rightarrow \neg p$), which may have helped to write it down in words.

Correct contrapositive of the initial implication $p \Rightarrow q$.

Working:

$U = 1, 2, 3, 4, 5, 6, 7, 8$

(a) (i) B = prime #'s <10 = $\{2, 3, 5, 7\}$

(ii) $\{2, 3, 5, 7\}$

$C = \{2, 3, 4, 5, 6\}$

$C \cap B = \{2, 3, 5\}$

(iii) $C' = \{1, 7, 8, 9\}$

$B = \{2, 3, 5, 7\}$

$B \cup C' = \{1, 2, 3, 5, 7, 8, 9\}$

(b) Contra: If x is not a positive integer btwn 1 and 7 then x is not a prime # <10

Contra positive $\neg q \rightarrow \neg p$

Answers:

(a) (i) $\{2, 3, 5, 7\}$

(ii) $\{2, 3, 5\}$ C4

(iii) $\{1, 2, 3, 5, 7, 8, 9\}$

(b) If x is not a positive integer between 1 and 7 then x is not a prime number less than 10   C2

Examiner report

Some students have difficulty identifying the elements of a given set, determining the elements of a complement of a set, and determining the elements of the union and the intersection of two sets. Students who hold misconceptions about an intersection of two sets often list more elements in the intersection than the elements in either of the two sets.

Similarly, students with misconceptions about a union of two sets often list fewer elements in the union than in each of the original sets. This question requires that students be able to use set notation and be familiar with the key concepts listed at the start of this chapter.

Part (b) of the question calls for the construction of the contrapositive of a given statement, which many students had difficulties with. Students should know that the contrapositive of the sentence "If p, then q" is "If not q, then not p". For example, the contrapositive of the sentence "If I visit Argentina, then I visit South America" is "If I don't visit South America, then I don't visit Argentina".

1. *[Maximum mark: 20]*

(i) When Geraldine travels to work she can travel either by car (*C*), bus (*B*) or train (*T*). She travels by car on one day in five. She uses the bus 50% of the time. The probabilities of her being late (*L*) when travelling by car, bus or train are 0.05, 0.12 and 0.08 respectively.

(a) Copy the tree diagram below and fill in all the probabilities, where *NL* represents not late, to represent this information. *[5 marks]*

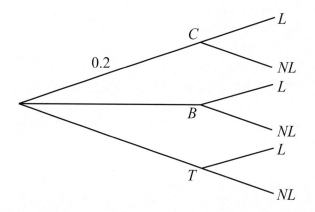

(b) Find the probability that Geraldine travels by bus and is late. *[1 mark]*

(c) Find the probability that Geraldine is late. *[3 marks]*

(d) Find the probability that Geraldine travelled by train, given that she is late. *[3 marks]*

[Taken from paper 2 May 2007]

How do I approach the question?

You should be able to understand that "She uses the bus 50% of the time" means that the probability of her travelling by bus is $\frac{1}{2}$, and that "She travels by car on one day in five" means that the probability that she travels by car is $\frac{1}{5}$. Since travelling by bus, car and train represent all of the possible ways that Geraldine can go to work, then the sum of the probabilities of Geraldine travelling by bus, car and train is 1. Thus the probability that Geraldine travels by train is

$1 - \frac{1}{2} - \frac{1}{5} = \frac{3}{10}$

Additionally, in part (a) you need to be able to recognize complementary events, for example, Geraldine is late, and Geraldine is not late, and infer that

$P(\text{Geraldine is not late}) = 1 - P(\text{Geraldine is late})$

For example, since we know that, when Geraldine travels by car, the probability of her being late is 0.05, then we can conclude that the probability of her not being late when she travels by car is

$1 - 0.05 = 0.95$

Similarly, we complete the rest of the tree diagram.

In (b) you have to determine the set of branches that leads to obtaining the outcome of Geraldine travelling by bus and being late. The probability of obtaining this result is calculated by multiplying the probabilities along the path taken, that is

$0.5 \times 0.12 = 0.06$

For (c) you need to determine all the paths following the tree diagram that Geraldine can take in order to arrive late at work. There are three different paths that lead to such an outcome. Hence we have to find the sum of all the probabilities along the paths taken.

And finally in (d) we have to calculate the probability that Geraldine travelled by train knowing that she was late, where we will use the formula for conditional probability.

What are the key areas from the syllabus?

- Representing alternative events by using tree diagrams
- Probability of events
- Conditional probability

This answer achieved 0/12

All of the probabilities calculated by the student are incorrect. The student's writing in the upper right corner of the working space indicates a misinterpretation of the information given in the question. For example, the student writes "Bus = 50% of $\frac{4}{5}$ = 0.4", which is an incorrect interpretation. It should have been understood that Geraldine travels by bus 50% of the time, and thus the probability of her taking the train is $\frac{1}{2}$. Similarly, the student's interpretation "Train = 50% of $\frac{4}{5}$ = 0.4" is incorrect. In addition, the student fails to recognize that the sum of the probabilities of "Geraldine is late" and "Geraldine is not late" is 1, since the two events are complementary.

The student identifies the path in the diagram that leads to the outcome Geraldine travelling by train and being late, but instead of multiplying the probabilities along this path, the student adds them.

The probability of Geraldine being late could be obtained from the diagram by identifying the paths that lead to this outcome and finding the sum of the products of the probabilities along those paths. The student fails to identify these paths. The use of 1.25, which represents a probability, is worrisome, as it shows that the student does not know that a probability of an event ranges from 0 to 1.

The student adds the probabilities of Geraldine being late in the three different cases when she travels by car, bus and train, which is an incorrect approach to answering the question. The student fails to recognize that the question concerns the calculation of conditional probability.

Working:

1 (i) (a)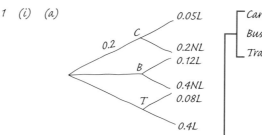

Car = $\frac{1}{5}$ = 0.2
Bus = 50% of $\frac{4}{5}$ = 0.4
Train = 50% of $\frac{4}{5}$ = 0.4   A0

(b) 0.12 + 0.4
= 0.52
∴ $P(A) = \frac{0.12}{0.52}$ M0

(c) (0.05 + 0.12 + 0.08) + (0.2 + 0.4 + 0.4)
= 0.25 + 1   A0
= 1.25
$P(A) = \frac{0.25}{1.25}$ M0 A0

(d) (0.05 + 0.12 + 0.08)
= 0.25   M0
∴ $\frac{0.4}{0.25}$ A0
= 1.6

This answer achieved 7/12

Working:

1 (i) (a)

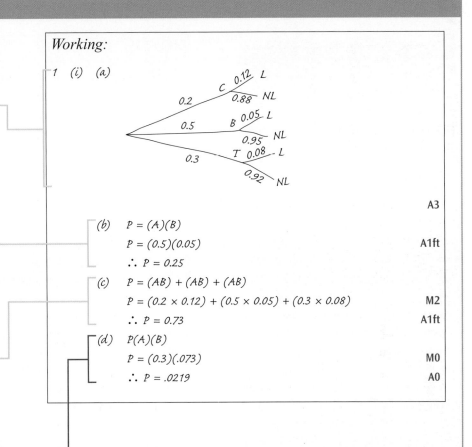

A3

(b) $P = (A)(B)$
   $P = (0.5)(0.05)$
   $\therefore P = 0.25$ A1ft

(c) $P = (AB) + (AB) + (AB)$
   $P = (0.2 \times 0.12) + (0.5 \times 0.05) + (0.3 \times 0.08)$ M2
   $\therefore P = 0.73$ A1ft

(d) $P(A)(B)$
   $P = (0.3)(.073)$ M0
   $\therefore P = .0219$ A0

This answer achieved 12/12

Full marks are awarded for this question. The student has successfully added their calculations, where the student explains the derivation of some of the entries, to the tree diagram.

The student identifies the path in the diagram that leads to the outcome that Geraldine travels by train and is late. Then the student correctly multiplies the probabilities along this path.

The probability of Geraldine being late is obtained from the diagram. The paths that lead to this outcome are identified, and then the sum of the products of the probabilities along those paths is calculated.

Correct choice of the formula for conditional probability and correct calculation. As with the student's previous answers, there is a clear indication of how the individual answers are derived from the formulae used, which makes it easy to follow!

Working:

1 (i) (a)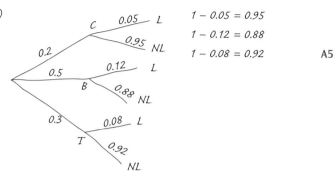

$1 - 0.05 = 0.95$
$1 - 0.12 = 0.88$
$1 - 0.08 = 0.92$ **A5**

(b) $P(\text{bus, late}) = 0.5 \times 0.12$
$= 0.06$ **A1**

(c) $P(\text{late}) = 0.06 + (0.2 \times 0.05) + (0.3 \times 0.08)$ **M2**
$= 0.094$ **A1**

(d) $P(\text{train}/\text{late}) = \dfrac{0.3 \times 0.08}{0.094}$
$= 0.25531$ **M1 A2**
≈ 0.255 (3 sig fig)

Examiner report

A common mistake with this type of question is the inability to interpret the information that is given in the question. For example, the problem says: "She travels by car on one day in five. She uses the bus 50% of the time." This is highlighted within the first student answer (0/12) as "Car = $\frac{1}{5}$ = 0.2" and "Bus = 50% of $\frac{4}{5}$ = 0.4", which shows that the student interprets it as "Geraldine uses the bus 50% of the time during which she does not use the car". This is an incorrect interpretation. You should read the question carefully and try to make sense of the information first, and then try to visualize it by using, for example, tree diagrams, Venn diagrams or tables. This question also requires knowledge of the probability of mutually exclusive events, combined probability and conditional probability.

7. Functions

What is a function?

You should be able to:
- recognize a function as a mapping of a set of *x* values onto a set of *y* values
- draw a mapping diagram.

You should know:
- domain (the allowed *x* values)
- range (the set of *y* values onto which the *x* values are mapped)
- $\mathbb{N}, \mathbb{Z}, \mathbb{Q}, \mathbb{R}$ notation.

Example

If $f: x \mapsto 2x - 1$ for $x \in \{1, 2, 3, 4, 5\}$:

(a) State the domain of *f*.

  Domain = {1,2,3,4,5,}

  The domain is the set of allowed x values. So in this example the domain is the set of values given in the question. If a domain is not given in a question, the domain is assumed to be $\mathbb{R}$, all real numbers.

  $f: x \mapsto 2x - 1$ *describes the function and how the range is obtained from the domain. f is the name of the function. Other ways of writing this function are $f(x) = 2x - 1$ or $y = 2x - 1$.*

(b) Draw a mapping diagram to represent the relationship between the domain and range.

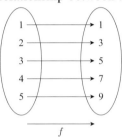

A mapping diagram shows the domain and range, in two distinct groups or sets, and the relationship between them. In this example, the relationship between the domain and range is that the x values need to be multiplied by 2, then 1 is subtracted from the product.

For example, if the value x = 3 is chosen from the domain, the corresponding value in the range is $(2 \times 3) - 1 = 5$. An arrow is then drawn from the value x = 3 in the first set, the domain, to a value that you calculate and write into the second set, the range.

(c) State the range of *f*.

  Range = {1, 3, 5, 7, 9}

  When you substitute each of the values of the domain into the expression 2x − 1, the values 1, 3, 5, 7, 9 are obtained. This set of numbers is called the range.

7. Functions

Linear functions

You should be able to:

- recognize the function notation form for linear functions $f : x \mapsto mx + c$
- sketch the graph of a line from an equation
- use your GDC to graph linear functions, and to find gradients and intercepts on axes

Texas Instruments

y=2x-1, Graph, CALC – Zero, CALC – dy/dx –at x=1

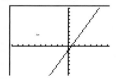
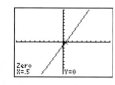
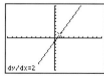

Casio

y=2x-1, DRAW – G-Solve for x-calc, Sketch – Tang – x=1 to get dy/dx

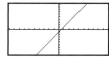
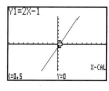

- solve real-world problems.

You should know:

- the equation of a line in both forms, $y = mx + c$ and $ax + by + d = 0$
- the gradient of a line between two points, $m = \dfrac{y_2 - y_1}{x_2 - x_1}$.

This topic links with "Coordinates in two dimensions" in chapter 8, where you are reminded how to plot points on a set of coordinate axes, find the mid-point of a line segment, and find the distance between two points.

Example

A taxi company in Darwin has a booking fee (fixed rate) of 5.50 AUD (Australian dollars) and charges 0.85 AUD per kilometre travelled.

(a) Write down a rule for the cost (C) of hiring a taxi in the Darwin area for each kilometre (d) travelled.

$C = 5.50 + 0.85d$

The cost of the taxi is made up of two parts: a fixed amount (the booking fee) and a variable part, which depends on how far you travel in the taxi.

(b) If the distance from the city to the airport is 13 km, find the cost of hiring a taxi for this trip, to the nearest Australian dollar.

$C = 5.50 + 0.85 \times 13$

$C = 5.50 + 11.05$

$C = \$16.55$

Cost is $17 to nearest dollar.

To calculate how much a taxi journey of 13 km will cost, you need to substitute the value x = 13 into the equation. Remember to add the $5.50, as this amount must be paid irrespective of the distance travelled.

(c) If a taxi fare is 52.80 AUD, calculate the distance travelled.

$52.80 = 5.50 + 0.85d$

$52.80 - 5.50 = 0.85d$

$47.30 = 0.85d$

$\dfrac{47.30}{0.85} = d$

$d = 55.6 \, km$

This question is asking the reverse of part (b). The taxi fare is known but the distance travelled needs to be calculated. Some rearranging of the equation must be done to find the distance travelled. Note that, if not otherwise stated, answers are given exactly or to three significant figures.

Continued

Linear functions (continued)

If you are using your GDC to calculate the answer to part (c), you could use 2nd TRACE (CALC) to find the point(s) of intersection of your cost function $C = 5.50 + 0.85d$ and the horizontal line $C = 52.80$.

Texas Instruments

$Y_1 = 5.5 + 0.85x$, $Y_2 = 52.8$

CALC – Intersection

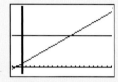 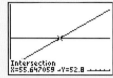

Casio

$Y_1 = 5.5 + 0.85x$, $Y_2 = 52.8$

G-Solve – ISCT

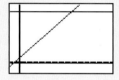 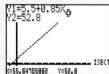

Quadratic functions

You should be able to:
- sketch the graph of a quadratic function from the equation
- find the y intercept of a parabola (the graph of a quadratic function)
- factorize quadratic equations to find the x intercepts of the graph of the function
- find the axis of symmetry and vertex of the parabola.

This topic links with "Solution of quadratic equations" in chapter 5.

You should know:
- standard form, $f : x \mapsto ax^2 + bx + c$
- the formula for the axis of symmetry from the equation in standard form, $x = -\dfrac{b}{2a}$
- that to find the x intercepts of a parabola with quadratic equation $y = 2x^2 - 5x - 3$, you need to solve the equation $2x^2 - 5x - 3 = 0$.

Texas Instruments
APPS – POLYSMLT

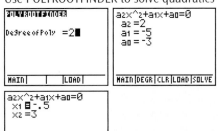

Use POLYROOTFINDER to solve quadratics

Casio
MENU – EQUA – F2-2

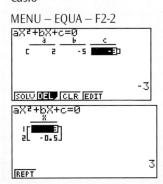

Example
The graph of the function $f(x) = 30x - 5x^2$ is given in the diagram below.

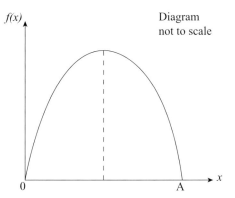

Diagram not to scale

(a) Factorize fully $30x - 5x^2$.

   $5x(6 - x)$

   Always look for a common factor before trying more complicated methods. The common factor in these two terms is x. The question asks you to factorize fully an expression. This is different from solving an equation. When you are asked to factorize, you find factors. Do not try to find values for x.

(b) Find the coordinates of the point A.

   $5x(6 - x) = 0$

   $x = 0$ or $x = 6$

   So A is the point (6, 0)

   In part (b) you need to recognize that you are trying to find an x intercept. One x intercept is at (0, 0) and the other is at A. To find where the graph cuts the x-axis, the y coordinate of the function equals zero. This is where you let $x(6 - x)$ equal zero to find the x coordinate of the intercept at A.

   Make sure you answer the question asked. In this question, you must find the coordinates of the point A. To achieve full marks, you must write down both the x and y coordinates of the point A.

Continued

Quadratic functions (continued)

(c) Write down the equation of the axis of symmetry.

Either $x = 3$

or $x = \dfrac{-b}{2a} = \dfrac{-30}{2 \times (-5)} = \dfrac{-30}{-10}$ *so* $x = 3$

The axis of symmetry can be found from the x intercepts. If you know the two x intercepts, then the axis of symmetry is half-way between the two intercepts. Half-way between $x = 0$ and $x = 6$ is $x = 3$.

The second method of finding the axis of symmetry is to use the equation. First, you need to identify a, b and c from the coefficients of each of the terms of the quadratic equation.

Remember to give the equation of the axis of symmetry as an equation, $x = 3$.

Casio

Use G-Solve to find intercepts

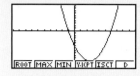
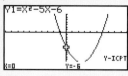

Press MIN to find turning point

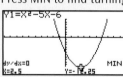

Be prepared

- Your GDC can help you to find accurate coordinates of the intercepts and vertex (turning point) of a parabola.
- It can also find values of y given the x value, zeros (x intercepts), maximum and minimum values and intersection points.

Texas Instruments

$Y = x^2 - 5x - 6$

Use CALC – Value to find y intercept

Use ZERO to find x intercepts

Use Minimum to find turning point

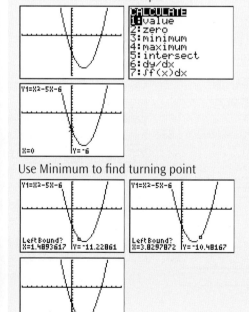

Exponential functions

You should be able to:

- recognize the standard form of an exponential function
- sketch graphs of the standard exponential function, $f(x) = a^x$, and its transformations
- apply exponential functions to growth and decay situations
- identify asymptotic behaviour of exponential functions.

This topic links with "Compound interest" in chapter 11.

Example

A bacteria culture grows such that the number of bacteria, N, present after t hours can be represented by

$$N = 2000 \times 4^{0.2t}$$

(a) Determine the number of bacteria present after 45 minutes. (Express your answer to the nearest whole bacterium.)

45 minutes = 0.75 hours

$$N = 2000 \times 4^{0.2 \times 0.75} = 2462.29$$

= 2462 (to the nearest whole number)

Always check the units of the question to make sure your substitutions are correct. In this question, the time was required in hours, and a whole number of bacteria is needed in the answer, so rounding is required.

(b) Calculate the length of time it takes for the number of bacteria to triple. (Give your answer to one decimal place).

$6000 = 2000 \times 4^{0.2t}$

Use your GDC to solve.

$t = 4.0 h$

Texas Instruments

MATH – Solver – enter equation – press alpha, enter to SOLVE

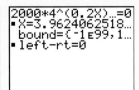

Casio

MENU – EQUA – enter equation – press solve

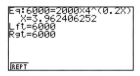

Be prepared

- Remember that a negative index as in $f(x) = 2^{-3x}$ indicates decay rather than growth. The function is decreasing.

Texas Instruments   **Casio**

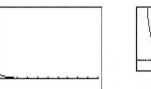

- $f(x) = \left(\frac{1}{2}\right)^x$ also indicates decay, as $\left(\frac{1}{2}\right)^x$ can also be written as 2^{-x}. If the base number is a positive fraction less than 1, and the index is positive, the function is also decreasing.

- When sketching exponential functions, always mark the position of any intercepts and asymptotes.

- Use your GDC to give you an idea of the shape of your exponential graph and the position of intercepts on axes and asymptotes.

Texas Instruments   **Casio**

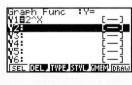

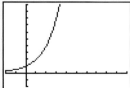
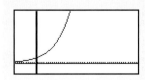

7. Functions

Trigonometric functions

You should be able to:
- recognize the standard form of sine and cosine functions
- sketch graphs of the standard sine and cosine functions and their transformations
- apply sine and cosine functions to real-life problems.

You should know:
- $f(x) = \sin x$ and $f(x) = \cos x$ have period 360° and amplitude 1.
- $f(x) = a \sin bx + c$ and $f(x) = a \cos bx + c$ have period $\dfrac{360}{b}$, amplitude a and vertical translation $+c$, where $a, b, c \in \mathbb{Q}$.

Be prepared
- Make sure your calculator is in degree mode.

Texas Instruments

MODE

Casio

MENU – RUN – SET-UP – SCROLL DOWN

Example

The depth of water, D metres, in a harbour, t hours after midnight, is given by

$D(t) = 4 + 2\sin 60t,\ 0 \leq t \leq 6$

(a) Write down the amplitude of the function.

Amplitude is 2 m

Texas Instruments

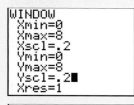

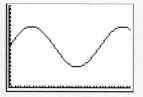

Casio

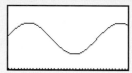

If you are given an equation, it is important to use your GDC to sketch the graph. From the graph, you can note all the important features.

(b) Determine the period of the function.

$p = \dfrac{360}{b} = \dfrac{360}{60} = 6h$

The period is found using the formula above or by looking at the graph. The period tells you how long it takes for the function to repeat itself.

(c) Sketch the graph of the depth for the first 6 hours.

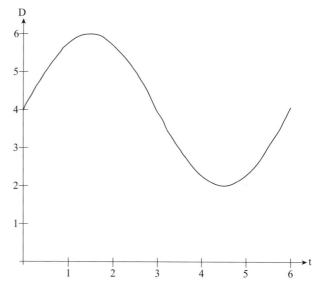

(d) Find the maximum and minimum water depths.

Maximum 6 m, minimum 2 m (these can be read from the graph).

Continued

Trigonometric functions (continued)

Be prepared

- Use your GDC to find *x* and *y* intercepts, to find maximum and minimum points, and to solve trigonometric equations of the form $3\sin 2x = 0.75$.

Texas Instruments

Casio

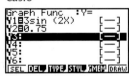

CALC – Intersection

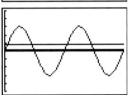

G-solve – ISCT

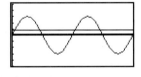

or

Texas Instruments

MATH – Solver

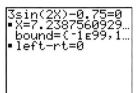

Casio

MENU – EQUA – EXE – Solver

- Make sure any graphs you sketch stay within the stated domain.
- Use ZoomFit to find a suitable scale for your graph.
- Lost a graph? This button finds it again.

Texas Instruments

ZoomFit

Casio

ZOOM – AUTO

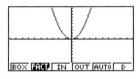

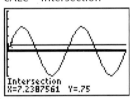

Other functions

You should be able to:

- use your GDC to sketch some functions that you may not have studied in depth
- sketch functions with equations such as $f(x) = \dfrac{1}{x+c}$ or $f(x) = x^3$
- locate intercepts on axes, maximum or minimum points, or asymptotes of these graphs
- solve equations (find points of intersection) of combinations of familiar and unfamiliar functions, using your GDC.

Texas Instruments

2nd TRACE (CALC)

Find the intersection of the lines $y = 3 - 2x$ and $y = 2^x$

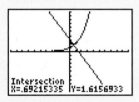

Casio

MENU – GRAPH – DRAW – G-SOLV

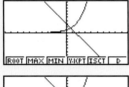

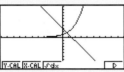

Find the intersection of the lines $y = 3 - 2x$ and $y = 2^x$

This gives one value. If there are more intersection points, just move the cursor to the right or left.

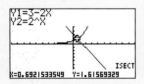

Be prepared

- If a question asks you to give an answer to a specific number of decimal places, you should immediately consider answering the question using your GDC.
- Examiners are aware of different forms of notation for writing intervals. So $-2 < y \leq 3$ and $(-2, 3]$ are both acceptable ways of expressing an interval.

11. (a) Sketch the graph of the function $y = 1 + \dfrac{\sin(2x)}{2}$ for $0 \leq x \leq 360°$ on the axes below.

[4 marks]

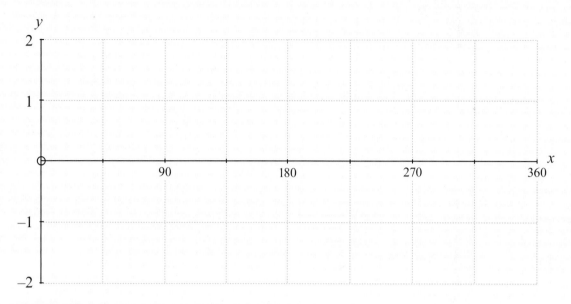

(b) Write down the period of the function. [1 mark]

(c) Write down the amplitude of the function. [1 mark]

[Taken from paper 1 November 2007]

How do I approach the question?

You have to think about the possible transformations of a sine function and which numbers affect its period and amplitude. If the function is written as $y = \frac{1}{2}\sin(2x) + 1$, then its amplitude and period are more obvious. The amplitude is equal to $\frac{1}{2}$, and is the maximum distance of the function from the equilibrium position. The period is found from the formula

$$p = \frac{360}{b}$$

where b is the number in front of the x term (or variable). In this question,

$$p = \frac{360}{2} = 180.$$

The function repeats itself every 180°.

When you are asked to sketch a graph, it is important to consider the x and y intercepts, the shape of your graph, and also the start and end points of the curve. The domain given in the question is important, so you know where to start and finish your sketch. In this question, the scale is provided for the student. Marks will be given for starting the graph in the appropriate place, a correct shape, and maximum and minimum points in the appropriate positions.

What are the key areas from the syllabus?

- Trigonometric functions including transformations
- The basic shape of a sine and cosine function

Which GDC functions will I need?

You will need to know how to sketch the graph of a sine function using your GDC. Make sure your mode is degrees, and that you have a suitable viewing window.

This answer achieved 1/6

The student has not adjusted the window on the calculator to 0 to 360°, resulting in a horizontal line for the answer to part (a).

The student achieved 1 mark for the correct y intercept/starting point, but missed the remaining 3 marks, which were awarded for correct minimum points, for correct maximum points and for drawing a smooth sine curve.

No follow-through marks have been awarded in part (b) for a period of 0, as the student's graph is significantly easier than the graph required. The student's graph does not have any amplitude. If the basic shape of a sine graph had been recalled, the student would have realized the graph was incorrect.

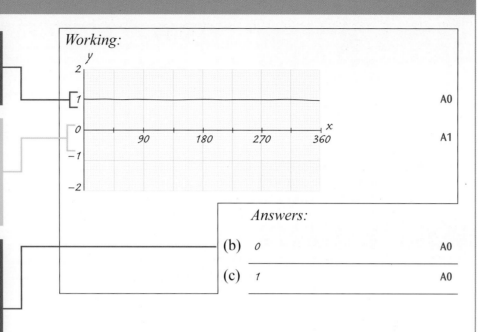

Answers:
(b) 0 — A0
(c) 1 — A0

This answer achieved 3/6

The student has the correct starting point (y intercept) on the graph and is awarded 1 mark for this. But the student has not been precise in copying the maximum and minimum points from the GDC. It is important to remember that marks are given for accuracy.

Both the period and the amplitude have been correctly stated. The period is found by noting where the graph starts to repeat itself (at 180°). The amplitude is found by halving the distance between the maximum and minimum values of the graph. These answers were awarded 1 mark each. If the student had incorrectly drawn the sine graph in part (a), but had given the correct values for the period and amplitude of the graph, follow-through marks would have been awarded here.

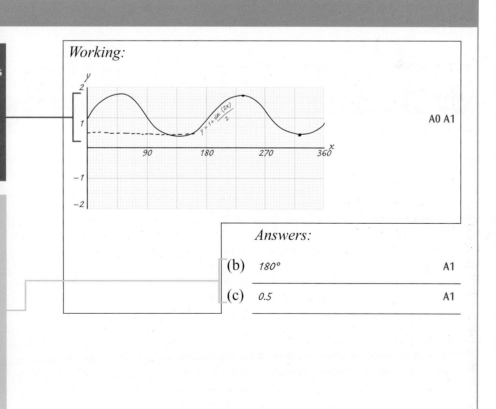

Answers:
(b) 180° — A1
(c) 0.5 — A1

This answer achieved 5/6

All the maximum and minimum points and the starting point have been accurately plotted by this student, and the curve is the correct shape.

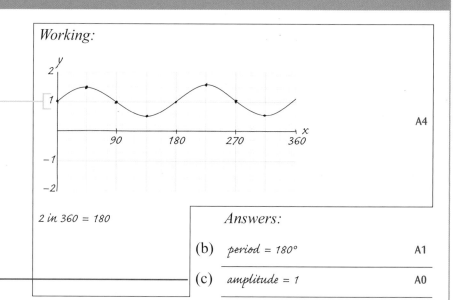

Working:

2 in 360 = 180

Answers:

(b) period = 180° A1

(c) amplitude = 1 A0

A4

The amplitude has been incorrectly found as the difference between the maximum and minimum value. This number should have been divided by 2.

Examiner report

This question assessed understanding of the transformations of a sine curve and the ability to sketch the curve using the GDC. Follow-through marks were awarded in parts (b) and (c) if an incorrect sine curve was drawn in part (a).

Common errors in this style of question include not having an appropriate viewing window, which means that an important part of the graph may not be observed on the screen, and not writing the function in standard form, which would make the period and amplitude more difficult to determine.

These problems can be overcome if the domain of the function is noted and the viewing window is set to match the domain. Remember to change the x scale to 30, 45 or 60, so you can read the position of the points you are locating, on the x-axis. The viewing window for y_{min} and y_{max} should then be chosen to make sure the whole graph is visible, including the maximum and minimum points. Alternatively, you can use ZoomFit from your ZOOM MEMORY.

12. (a) $f : x \rightarrow 3x - 5$ is a mapping from the set S to the set T as shown below.

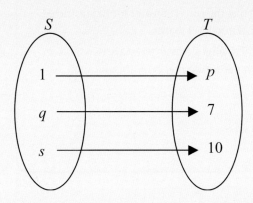

   Find the values of p and q. [2 marks]

(b) A function g is such that $g(x) = \dfrac{3}{(x-2)^2}$.

   (i) State the domain of the function $g(x)$. [2 marks]

   (ii) State the range of the function $g(x)$. [1 mark]

   (iii) Write down the equation of the vertical asymptote. [1 mark]

[Taken from paper 1 November 2007]

How do I approach the question?

(a) The relation from S to T is given by the function f. You need to work out when to substitute $x = 1$ to find p and when to substitute $f(x) = 7$ to find q.

(b) The function has a restricted domain. The denominator cannot equal zero or the function will be undefined. You have to remember that the domain represents the x values and the range represents the y values allowed by the function.

What are the key areas from the syllabus?

- Mappings
- Domains
- Ranges
- Identifying asymptotes

7. Functions

What is the correct notation?

While $\mathbb{R}, x \neq 2$, gives one method of writing the correct answer for the restricted domain, all correct alternatives will be awarded full marks.

When writing the equation of the vertical asymptote, it is necessary to write $x = 2$. Just writing an answer of 2 will not allow maximum marks for this question.

Which GDC functions will I need?

By selecting an appropriate window and graphing the function, the restrictions on the domain and range can be seen as vertical and horizontal asymptotes.

ZoomFit will be useful here, to allow you to find your graph if the window is making it look too big or too small.

This answer achieved 2/6

The student has correctly found p and q by solving the linear equations

Unfortunately, the student has not appreciated that parts (a) and (b) are unrelated. The question is to find the domain and range of a new function $g(x)$. It is important to read each question several times to make sure you have understood whether or not the parts of the question are connected.

The student appears to be confusing the graph of a quadratic function with the graph of this function. It is necessary to know, and to be able to identify, the graphs of each of the functions in the mathematical studies course.

Working:

$p = 3.1 - 5 = -2$

$q \quad 3x - 5 = 7$

$\quad 3x = 7 + 5$

$\quad x = \dfrac{12}{3}$

$\quad x = 4$

$\dfrac{-b}{2a}$

$\dfrac{5}{2.3} \qquad y = 3x - 5$

$\dfrac{5}{6} \qquad y = 2.5x$

Answers:

(a) $p = -2 \qquad q = 4$ A2

(b) (i) S A0

     (ii) T A0

     (iii) $\dfrac{-b}{2a}$ A0

This answer achieved 4/6

The student has made a mistake when doing the subtraction $p = 3 - 5$. The answer to this should be -2, but the student has written 2. The student receives no marks for this incorrect answer, but 1 mark for correctly finding $q = 4$.

In part (b), 2 marks were awarded for answering "all real numbers, except $x = 2$". This particular value of x will give a denominator of zero, and the function is undefined at $x = 2$. There are many equivalent notations and all correct ones are awarded marks. This student was awarded 2 marks for correctly identifying the domain of the function. The first answer mark was given for identifying $\mathbb{R}$, the set of real numbers, and the second answer mark was awarded for excluding the value $x = 2$.

The range is the set of allowed y values and is positive for this function. Stating the range as $g(x) > 0$, $y > 0$, $(0, \infty)$, $]0, \infty[$ or $\mathbb{R}^+$ would all have been appropriate answers.

Working:

$f : x \to 3x - 5$
$p = 3 \times - 5$
$p = 3 \times 1 - 5$
$p = 3 - 5$
$p = +2$
$f = 3q - 5$
$f + 5 = 3q$
$12 = 3q$
$\frac{12}{3} = q$
$4 = q$

Answers:

(a) $p = +2$ and $q = 4$ A0 A1

(b) (i) $\text{dom} = \mathbb{R} - \{2\}$ A1 A1

(ii) $\text{range} = \mathbb{R}$ A0

(iii) $x = 2$ A1

This answer achieved 6/6

This student shows an alternative way of writing the domain. Remember that round brackets, a strict equality sign or $\infty[$ are always used for infinity.

The student has correctly identified the domain, including the value of x where the function is undefined. The range is correctly identified as positive real numbers, and the student achieves the maximum possible score.

Working:

Answers:

(a) $p: -2$ $q = 4$ A2

(b) (i) $(-\infty, 2), (2, \infty)$ A2

(ii) $(0, \infty)$ A1

(iii) $x = 2$ A1

Examiner report

Whenever the equation of a graph is provided in a question, you should sketch the graph on your GDC to see if it is one you recognize and check its behaviour. Sketching the graph enables you to see intercepts, asymptotes and maximum or minimum points.

Common errors in this style of question include mistakes in arithmetic when substituting values. Try not to skip steps, as this can sometimes lead to an incorrect answer to what is otherwise a simple problem. You should read the whole question to see if there is a relationship between the different parts and whether information or answers in one part can be used in subsequent parts of the question.

Remember that vertical asymptotes occur when a function has a denominator containing x. In this question, the denominator is $(x - 2)^2$. Since $x - 2$ cannot equal zero, $x \neq 2$, so the equation $x = 2$ is a vertical asymptote. The graph does not go through this line, but approaches it from either side.

3. *[Maximum mark: 17]*

(i) The following graph shows the temperature in degrees Celsius of Robert's cup of coffee, t minutes after pouring it out. The equation of the cooling graph is $f(t) = 16 + 74 \times 2.8^{-0.2t}$ where $f(t)$ is the temperature and t is the time in minutes after pouring the coffee out.

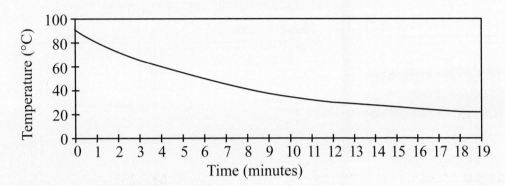

(a) Find the initial temperature of the coffee. *[1 mark]*

(b) Write down the equation of the horizontal asymptote. *[1 mark]*

(c) Find the room temperature. *[1 mark]*

(d) Find the temperature of the coffee after 10 minutes. *[1 mark]*

If the coffee is not hot enough it is reheated in a microwave oven. The liquid increases in temperature according to the formula

$$T = A \times 2^{1.5t}$$

where T is the final temperature of the liquid, A is the initial temperature of coffee in the microwave and t is the time in minutes after switching the microwave on.

(e) Find the temperature of Robert's coffee after being heated in the microwave for 30 **seconds** after it has reached the temperature in part (d). *[3 marks]*

(f) Calculate the length of time it would take a similar cup of coffee, initially at 20°C, to be heated in the microwave to reach 100°C. *[4 marks]*

[Taken from paper 2 November 2007]

How do I approach the question?

The intercepts, shape and any asymptotes on the graph will provide useful information to help you to answer the questions.

What are the key areas from the syllabus?

- Graphing techniques
- Exponential functions, including increasing and decreasing functions

Which GDC functions will I need?

You will need to know how to sketch the graph of an exponential function, changing the viewing window in the y direction (for the temperature) to account for the vertical translation of 16 and the dilation factor of 74. Make sure, when you are entering -0.2, that you use the negative sign $(-)$ rather than the subtraction sign on your GDC. Remember that you must always use mathematical notation, not calculator notation. For example, write x^2 not x^2.

Texas Instruments

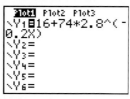

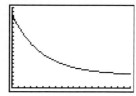

Casio

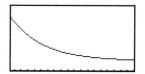

7. Functions

This answer achieved 3/11

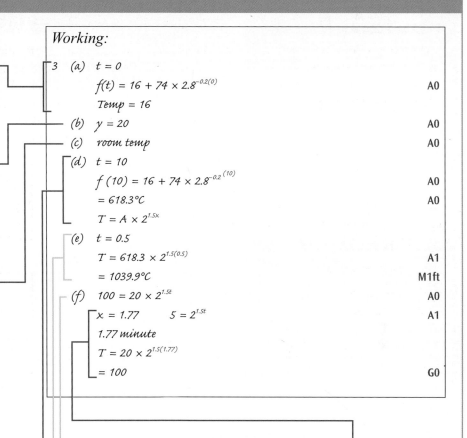

Working:

3. (a) $t = 0$
   $f(t) = 16 + 74 \times 2.8^{-0.2(0)}$ A0
   Temp = 16

(b) $y = 20$ A0

(c) room temp   A0

(d) $t = 10$
   $f(10) = 16 + 74 \times 2.8^{-0.2(10)}$ A0
   $= 618.3°C$ A0
   $T = A \times 2^{1.5x}$

(e) $t = 0.5$
   $T = 618.3 \times 2^{1.5(0.5)}$ A1
   $= 1039.9°C$ M1ft

(f) $100 = 20 \times 2^{1.5t}$ A0
   $x = 1.77$ $5 = 2^{1.5t}$ A1
   1.77 minute
   $T = 20 \times 2^{1.5(1.77)}$
   $= 100$ G0

Comments:

- This student has mistaken 74×2.8^0 to be equal to zero, then added 16 to 0 to reach a final answer of 16. Although $2.8^0 = 1$, when this is multiplied by 74, the answer is 74. Adding 16 to 74 would give the correct answer of 90°.

- The student has tried to find the asymptote from the graph, rather than more accurately from the equation. The *order* of operations is also important. Remember to multiply or divide before adding and subtracting.

- The question asked the student to find the room temperature, and the student simply wrote down the words *room temp*, without answering the question. A guess would have been a better alternative here, and may even have given this student an extra mark. Remember, marks are not deducted for incorrect answers, so if you have absolutely no idea how to do a question, or are running short of time, *guess*.

- A mistake here leads to an answer that could not be correct. This is a cup of coffee that is cooling down from a temperature of 90°. After 10 minutes, the coffee could not possibly be at a temperature 618.3°C. Always check that the answer to your question makes sense in a real-world situation.

- The student has received a follow-through mark for using the correct method with an incorrect value found in the previous part of the question. The first mark is for converting 30 seconds to 0.5 minutes and substituting it into the equation $T = A \times 2^{1.5t}$. Although 618.3 was an incorrect answer in part (d), it is the answer to part (d) that must be used as the initial temperature of the coffee before it is reheated in the microwave. In part (e), a follow-through mark is awarded for this step.

- The student used $t = 0.5$ minutes, correctly converting the seconds to minutes, and 1 mark was awarded for showing that $20 \times 2^{1.5t} = 100$.

- The final 3 marks were not awarded, as the answer given was incorrect and no working was shown. It is most important that questions worth 2 marks or more have working shown, particularly if you are not completely confident of your answer.

7. Functions

This answer achieved 6/11

The first answer mark was awarded for a correct answer to the initial temperature. Remember that *initial* means "at the start" or, in mathematical terms, when $t = 0$. Substituting $t = 0$ into the equation $f(0) = 16 + (74 \times 2.8^{-0.2(0)})$ gives an answer of 90, since $2.8^{-0.2(0)} = 1$:
$f(0) = 16 + (74 \times 1) = 16 + 74 = 90$
The student remembered to include units and was awarded 1 mark for a correct answer of 90°C.

The student did not show working when using the GDC. If using your calculator with no working, you need to be extremely confident that your answer is correct, otherwise you will lose valuable marks! Also, it is important to give the answer as an equation and not just as a number.

The student correctly substituted $t = 10$ into the equation to find the temperature of the coffee after 10 minutes. Since this question is only worth 1 mark, no working is required and the GDC would be used to find the correct answer of 25.4°C. The student has also correctly answered this question to three significant figures.

Working:

3 (a) (i) initial temp = 90°C   A1
  (b) horizontal asymptote = 16   A0
  (c)
  (d) $f(t) = 16 + 74 \times 2.8^{-0.2(10)}$ A1
       $t = 25.4°C$
  (e) $T = 25.4 \times 2^{1.5 \times .5}$ A1
       $T = 25.4 \times 2^{.75}$ M1
       $T = 42.7°C$ A1
  (f) $100°C = 20° \times 2^{1.5(t)}$ A1
       $0 = 20 \times 2^{1.5(t)} - 100$
       $= 1.76 \ min$ G0

62

This answer achieved 11/11

Substitutions of $t = 0$ and $t = 10$ were correctly made in parts (a) and (d) to achieve maximum marks.

Trying to read the equation of the horizontal asymptote from the graph given in the exam paper does not provide an accurate enough answer. To find the equation more accurately, the GDC can be used to zoom in on the appropriate region of the graph. The asymptote must be written as $y = 16$ to achieve the answer mark.

Parts (b) and (c) of this question were related. If you could find the position of the horizontal asymptote, you would probably notice that this was the temperature that the graph approached as time passed. This is another way of asking for the room temperature, as the coffee will eventually cool to the temperature of the room.

A new function is provided because the coffee is now being heated in a microwave oven. The initial temperature used in this part of the question is the answer to part (d) (25.4°C). The 30 seconds are converted to 0.5 minutes, then part of the required equation becomes $2^{1.5 \times 0.5}$. If this value is seen in the student's work, it is awarded 1 mark. A method mark is given for multiplying $2^{1.5 \times 0.5}$ by the student's answer to part (d). The final answer (42.7°C), using correct values, received 1 mark.

Working:

3 (i) $f(t) = 16 + 74 \times 2.8^{-0.2t}$

(a) $f(0) = 16 + 74 \times 2.8^{-0.2(0)}$
  $= 90°C$ **A1**

(b) $y = 16$ **A1**

(c) room temperature = 16°C **A1**

(d) after 10 minutes, coffee temp = 25.4°C **A1**

 $T = A \times 2^{1.5t}$ **A1**

(e) $T = 25.4 \times 2^{1.5 \times 0.5}$ **M1**
  $= 42.7°C$ **A1**

(f) $T = A \times 2^{1.5t}$ **A1**
  $100 = 20 \times 2^{1.5t}$ **M1**
  $0 = 20 \times 2^{1.5t} - 100$
  $t = 1.55$ **A2**
  ∴ it would take 1.55 minutes to reach 100°C

The temperature is given and the time taken for the coffee to reach a temperature of 100°C is to be found. To receive full marks for this question, the equation $100 = 20 \times 2^{1.5t}$ needs to be shown for the first answer mark, then the GDC can be used to find the solution. If a correct solution of 1.55 minutes or 92.9 seconds is given, the student is awarded 3 marks. If working is shown, with a correct answer, the student will receive further marks, a method mark for any correct method that leads to the correct answer, an accuracy mark for seeing a value of t between 1.54 and 1.56, and another accuracy mark for 1.55 minutes.

Examiner report

In this question, it is necessary to check that you are using the correct equation with appropriate units. Although seconds are given in one of the question parts, the equations for cooling and heating the coffee are given in terms of minutes. It is necessary to convert the seconds to minutes before substituting.

Common errors in this style of question include: trying to read the answers to questions from the graph rather than calculating the answers from the given equation; not writing the function in standard form, which would make the intercepts and vertical translations more difficult to determine; and not recognizing that, although the parts of the question are related, some parts require the use of a different equation with a different set of initial conditions.

These problems can be overcome if the question is read carefully more than once, and important points are highlighted. Sketching the graph(s) with an appropriate viewing window will help to identify important parts of the graph.

8. Geometry and trigonometry

Coordinates in two dimensions

You should be able to:
- plot points on the plane when the coordinates are given
- write down the coordinates of a point from the graph
- find the coordinates of the mid-point of a line segment
- find the distance between two points.

You should know:
- mid-point of a line segment with end points (x_1, y_1) and (x_2, y_2) is
$$\left(\frac{x_1 + x_2}{2}, \frac{y_1 + y_2}{2}\right)$$
- distance between points (x_1, y_1) and (x_2, y_2) is
$$\sqrt{(x_1 - x_2)^2 + (y_1 - y_2)^2}$$

Example

Points A(1, 3), B(−1, −2) and C are plotted in the following diagram.

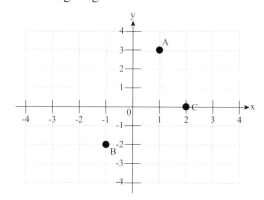

(a) Write down the coordinates of point C.

C(2, 0)

(b) Find the mid-point, M, between A and B.

Mid-point is at $M\left(\dfrac{1 + (-1)}{2}, \dfrac{3 + (-2)}{2}\right) = M\left(0, \dfrac{1}{2}\right)$

(c) Find the distance between A and B.

Let $x_1 = 1, y_1 = 3, x_2 = -1, y_2 = -2$. Then the distance d between A and B is

$d = \sqrt{[1 - (-1)]^2 + [3 - (-2)]^2} = \sqrt{2^2 + 5^2} = \sqrt{29}$

Be prepared
- When using the formula for distance, remember to use the brackets. Marks are often lost for omitting them.

Lines in two dimensions (2D geometry)

You should be able to:

- find the equation of a line
- draw a line when its equation is given
- find the gradient of a line and any intercepts of a line on the axes
- calculate the point of intersection of two lines
- determine whether two lines are parallel and whether two lines are perpendicular.

This topic links with "Solutions of a pair of simultaneous equations" in chapter 5.

You should know:

- the equation of a straight line (not vertical), $y = mx + c$, $ax + by + d = 0$
- the equation of a vertical line, $x = k$, where k is a constant
- the gradient of a line passing through (x_1, y_1) and (x_2, y_2) is $m = \dfrac{y_2 - y_1}{x_2 - x_1}$
- gradients of perpendicular lines are negative reciprocal, $m_1 = \dfrac{-1}{m_2}$, with $m_2 \neq 0$.

Example

The diagram shows the line L_1 passing through A(3, 3) and B(9, 6).

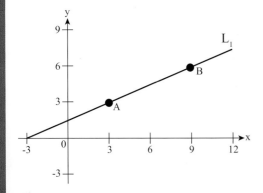

(a) Find the equation of L_1.

Equation of line is $y = mx + c$, where

$$m = \frac{y_2 - y_1}{x_2 - x_1} = \frac{6 - 3}{9 - 3} = 0.5$$

So far $y = 0.5x + c$

We still need to find c. Replacing the coordinates of either A or B in this equation, we find c.

$6 = 0.5 \times 9 + c$, *then $c = 1.5$*

So the equation of L_1 is $y = 0.5x + 1.5$

(b) Write down the coordinates of the point of intersection between L_1 and the y-axis.

Put $x = 0$ into the equation of the line L_1:

$y = 0.5 \times 0 + 1.5 = 1.5$

The point is (0, 1.5).

(c) A second line, L_2, is perpendicular to L_1. Write down its gradient.

$m_2 = \dfrac{-1}{0.5} = -2$

(d) Determine whether the line L_3 given by $x - 2y - 10 = 0$ is parallel to L_1.

Parallel lines have equal gradients. Find the gradient of L_3 and see if it is 0.5.

Rearranging the equation of L_3 to make y the subject of the formula gives

$2y = x - 10$

$y = 0.5x - 5$

Gradient of L_3 is 0.5. The two lines have the same gradient. They are parallel.

Be prepared

- To find the intersection with the x-axis, replace y by 0 in the equation of the line and find x.
- If the equation is $y = mx + c$, then the y intercept is c and the point is $(0, c)$.
- Vertical lines have no gradient.

Right-angled trigonometry

You should be able to:
- draw a clear and labelled diagram to represent the information given in a problem
- use Pythagoras' theorem and the ratios $\sin \alpha$, $\cos \alpha$ and $\tan \alpha$ to find unknowns in right-angled triangles.

You should know:
- Pythagoras' theorem states that in a **right-angled triangle** the square of the length of the hypotenuse side (a), is equal to the sum of the squares of the other two sides

- Pythagoras' theorem, $b^2 + c^2 = a^2$
- $\sin B = \frac{b}{a}$, $\cos B = \frac{c}{a}$, $\tan B = \frac{b}{c}$

Example
In the triangle EDF, EF = 8 cm, D = 50° and E = 90°.

(a) Draw and label a clear diagram to represent the information given.

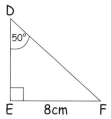

(b) Calculate the length of DE.

The trigonometric ratio that links DE and both the angle D and EF is the tangent. So

$\tan D = \frac{EF}{DE}$

$\tan 50° = \frac{8}{DE}$

$DE = \frac{8}{\tan 50°} = 6.71 \, cm$

Be prepared
- Your GDC must always be set in **degrees**, otherwise you will not get the correct answer.
- One common way to remember the trigonometric ratios listed above is through the phrase **SOHCAHTOA**.

Non-right-angled triangles

You should be able to:
- draw a clear and labelled diagram to represent the information given in a problem
- use the sine and cosine rules to find unknowns in triangles
- calculate the area of a triangle.

You should know:
- the **sine rule** is used when you are given
  - two sides of a triangle and a non-included angle or
  - two angles and one side
- the **cosine rule** is used when you are given
  - two sides and the included angle or
  - three sides of a triangle
- sine rule,
$\frac{a}{\sin A} = \frac{b}{\sin B} = \frac{c}{\sin C}$
- cosine rule,
$a^2 = b^2 + c^2 - 2bc \cos A$ or $\cos A = \frac{b^2 + c^2 - a^2}{2bc}$
- area of a triangle = $\frac{1}{2}ab \sin C$, where a and b are adjacent sides, and C is the included angle.

Example
A large field has a triangular shape as shown in the diagram.

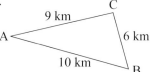

(a) Find the size of angle B. Give your answer correct to two decimal places.

Use the cosine rule in the form

$\cos B = \frac{a^2 + c^2 - b^2}{2ac}$

$= \frac{6^2 + 10^2 - 9^2}{2 \times 6 \times 10}$

$= \frac{55}{120}$

$B = 62.72°$ (2 d.p.)

(b) Find the area of the field.

To use the formula for the area of the triangle, you need two sides and the included angle. Using angle B just calculated and sides AB and BC gives

area = $\frac{1}{2} \times 10 \times 6 \times \sin 62.72°$

$= 26.7 \, km^2$

Geometry of three-dimensional shapes (3D geometry)

You should be able to:
- find surface area and volume of cuboids, prisms, pyramids, cylinders, spheres, hemispheres and cones
- calculate lengths of lines joining vertices with vertices, vertices with mid-points, and mid-points with mid-points
- identify angles between two lines, and between lines and planes
- find the size of angles between two lines, and between lines and planes.

You should know:
- in the information formula booklet you have a list of formulae for volume and surface area of all the solids listed above—the most important thing is to clearly identify the solid involved in the question
- sometimes the solid given in the question is a combination of two of the solids given above (imagine a cylinder with a hemisphere at the top)—in these cases you need to be careful with the way these formulae are used.

Example

The dimensions of an open box are as shown in the diagram.

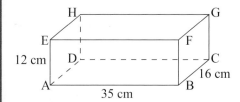

(a) Find the total surface area.

Add the area of the five faces of the open box.

Total surface area is 1784 cm²

(b) Find the distance between A and C.

Apply Pythagoras' theorem in the right-angled triangle ABC.

Distance between A and C = $\sqrt{1481}$ = 38.5 cm

(c) Find the angle between ABCD and the line AG.

The angle is GAC (see diagram). Notice that triangle GAC is right-angled at C.

$\tan GAC = \dfrac{12}{\sqrt{1481}}$

$GAC = 17.3°$

Be prepared
- When you are asked to find angles between lines, and between lines and planes, try to identify a right-angled triangle, which will lead you to the answer.
- When you are asked to find the surface area of a solid, the first thing to do is to count the number of faces.

15. The length of one side of a rectangle is 2 cm longer than its width.

   (a) If the smaller side is x cm, find the perimeter of the rectangle in terms of x.

   The perimeter of a square is equal to the perimeter of the rectangle in part (a).

   (b) Determine the length of each side of the square in terms of x.

   The sum of the areas of the rectangle and the square is $2x^2 + 4x + 1$ (cm²).

   (c) (i) Given that this sum is 49 cm², find x.

   (ii) Find the area of the square.

[Taken from paper 1 May 2007]

How do I approach the question?

A good way to approach a question like this is to make a diagram and put the given information in it. This helps to understand the situation much better.

(a) In the first part, the diagram would be:

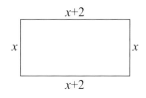

This makes it is easier to find the expression for the perimeter of the rectangle.

(b) Once you have found this, you need to find an expression for the side of a square with the same perimeter.

(c) The last part can be done regardless of the first two parts. In (i) you have to solve a quadratic equation, for which you can use your GDC.

What presumed knowledge should I have?

You should know basic manipulation of simple algebraic expressions, evaluating formula by substitution, and the formulae for the perimeter of a rectangle and the area of a square.

What are the key areas from the syllabus?

Solution of quadratic equations by use of the GDC or by use of the formula, which can be found in the information formula booklet.

Texas Instruments

MATH – SOLVER

```
EQUATION SOLVER
eqn:0=2X²+4X-48
```
```
2X²+4X-48=0
 •X=4
  bound={-1E99,1…
 •left-rt=0
```

Casio

MENU – EQUA – SOLVER

```
Eq:2X²+4X+1=49
  X=4
  Lft=49
  Rgt=49

REPT
```

This answer achieved 2/6

The student made a little diagram and put in the expressions for the length and for the width. The expression for the perimeter is correct.

The answer is correct because the side of the square is found by dividing the perimeter by 4.

It seems that the student did not understand the question. Look at the diagram made in (a) and notice that the 49 cm² is written in the rectangle. This is a clear misunderstanding of the question, which mentions the "sum of the areas of the rectangle and the square" and says that "this sum is 49 cm²".

In (c)(ii) the answer does not follow through from (c)(i), as $x + 1$ is just the side of the square and not its area!

Working:

(a) [rectangle: 49 cm², width x, length $x + 2$ cm]

(b) $\dfrac{4x + 4 \text{ cm}}{4} = x + 1 \text{ cm}$

(c) [sketch of parabola] $2(x + 0.293)(x + 1.71)$

$2(1.71)^2 + (1.71) \times 4 + 1 = 13.68$

(c) (ii) $(13.68) + 1 = 14.68$

Answers:

(a) $4x + 4$ cm — A1

(b) $x + 1$ cm — A1

(c) (i) $x = 13.7$ — A0

(ii) 14.7 cm² — A0

This answer achieved 4/6

The student did make a diagram but did not write the correct algebraic expression for the longer side and therefore found the wrong expression for the perimeter.

The quadratic equation is written and the correct solution is given.

The expression written here is consistent with the answer given in (a) (hence the follow-through), but the units are missing. Therefore, no mark was gained due to the unit penalty.

This answer follows through from the expression given for the length of the side of the square in part (b).

Working:

[rectangle: width x, length $2x$] [square: $1.5x \times 1.5x$]

$2x + 2x + x + x = 6x$

$2x^2 + 4x + 1$ $2x^2 + 4x + 1 = 49$

Answers:

(a) $6x$ — A0

(b) $1.5x$ — A0ft (UP)

(c) (i) 4 — C3

(ii) 36 cm² — A1ft

This answer achieved 5/6

This is a good diagram. It shows that the student fully understands the question. The rectangle was drawn with its side lengths, x and $x + 2$, written correctly.

The student had the correct answer for the perimeter, though the unit (cm) is missing, for which a unit penalty was applied.

Both the equation set to find x and the value of x are correct.

Although the student has done well to show the working, if we look at the working we can see an algebraic mistake. Have you found where it is? The student copies the equation $2x^2 + 4x + 1 = 49$, which is equivalent to writing $2x^2 + 4x = 48$. Dividing this equation by 2, you get $x^2 + 2x = 24$, and **not** $x^2 + 4x = 24$. However, the student does not lose any marks because the correct answer, $x = 4$, is written down.

The student works out the correct area by adding 1 to x and squaring the result. Notice that the student was not penalized here for not writing down the units (cm²) because the unit penalty was already deducted in a previous part.

Working:

x x

$x + 2$

$\dfrac{(x + 2) + (x + 2) + x + x}{2}$

$2x^2 + 4x + 1 = 49$

$\dfrac{48}{2} = x^2 + 4x$

$24 = x^2 + 4x$

Answers:

(a)	$4x + 4$		A0 UP
(b)	$x + 1$		A1
(c)	(i)	$x = 4$	C3
	(ii)	$5^2 = 25$	C1

Examiner report

This was the last question of the paper, which is often one of the hardest. This question was a combination of algebra and geometry (presumed knowledge), as it deals with expressions for the perimeter of rectangles and squares and also with their areas. Students generally understood parts (a) and (b) quite well, as these parts involve just "pre-syllabus knowledge", namely, the perimeter of a rectangle, and the basic manipulation of simple algebraic expressions. However, a common error is forgetting to write down the unit and losing 1 mark. It is worthwhile highlighting this type of information within the exam question so that you do not lose an easy mark. A common error within part (c) was students managing to write down the equation of the quadratic in part (i) but not using the GDC to solve it. Also, as x represents the side of a rectangle, it cannot be negative! Part (ii) is a difficult question. A common mistake was to take x as the length of the square, whereas the expression for the length of the square had been found in (b).

13. The mid-point, M, of the line joining $A(s, 8)$ to $B(-2, t)$ has coordinates $M(2, 3)$.

(a) Calculate the values of s and t. *[2 marks]*

(b) Find the equation of the straight line perpendicular to AB, passing through the point M. *[4 marks]*

[Taken from paper 1 November 2007]

How do I approach the question?

(a) The key word within this question is "mid-point". This leads you to the idea that you are definitely in geometry. You just need to use the formula, replacing the values from the problem in it.

(b) You need to find the equation of a line, so you are still in geometry. The line you are looking for is perpendicular to AB (with this information, you get the gradient) and passes through the mid-point M given in the problem (with this information, you get the y intercept). Sketching a diagram may help you to understand the situation.

How does this relate to the information formula booklet?

You need to use the formulae for:
- the mid-point of a line segment with given end points
- the gradient
- the equation of a straight line.

Is there any other formula that I need to use that is not in the information formula booklet?

You need to know that gradients of perpendicular lines are negative reciprocal:
$m_1 = \frac{-1}{m_2}$, with $m_2 \neq 0$.

8. Geometry and trigonometry

This answer achieved 2/6

Within the working, the student puts the coordinates of A and B in the formula for the mid-point, then finds the correct values for s and t, although it is not clear enough how the values were found.

It is not clear where the equation of the line comes from. Why write the formula for the vertex of a parabola, if the topic is straight lines? Since this is a geometry question, you just have to use the formulae available in the information formula booklet related to this topic. It would have been an improvement if the student had drawn a diagram with the line AB, the mid-point M and the perpendicular passing through M.

Working:

$\dfrac{x_1 + x_2}{2}; \quad \dfrac{y_1 + y_2}{2} \qquad (s; 8)$

$\dfrac{s + (-2)}{2}; \quad \dfrac{8 + t}{2} \qquad xv = \dfrac{-b}{2a}$

$4 - 2 \qquad 8 + -6 \qquad y = Mx + c$

$6 - 2$

$\dfrac{4}{2} \qquad\qquad 13.28$

2

$y = \dfrac{2}{3}$

Answers:

(a) $s = 6 \quad t = -2$ — A2

(b) $y = \dfrac{2}{3}x + 3$ — A0

This answer achieved 3/6

Within the working, we can see that the student puts the coordinates of A and B in the formula for the mid-point, then finds the values for s and t by setting two equations. Therefore, the student gets the correct answer in (a) and full marks.

The student finds the gradient of line AB and then makes a mistake when finding the gradient of the perpendicular line to AB. Remember that perpendicular lines have opposite and reciprocal gradients, which would have given the correct answer.

Working:

(a)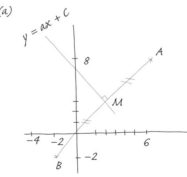

$\text{Midpoint} = \left(\dfrac{x_1 + x_2}{2}, \dfrac{y_1 + y_2}{2}\right)$

$\text{Midpoint} = (2, 3)$

$\dfrac{s - 2}{2} = 2$

$s - 2 = 4$

$s = 6$

$\dfrac{8 + t}{2} = 3$

$8 + t = 6$

$t = -2$

$\text{gradient } AB = \dfrac{-2 - 8}{-2 - 6}$

$= \dfrac{10}{8}$ — A1

(b) $y = ax + b$

gradient $= \dfrac{8}{10}$

$y = \dfrac{8}{10}x + $ — A0

Answers:

(a) $s = 6; t = -2$ — C2

(b) _____

This answer achieved 6/6

There is correct and clear working, with two equations set (one for s and one for t) and solved, one for each coordinate.

Clear reasoning is shown. First of all, the student finds the gradient of the line AB, then the gradient of a line perpendicular to AB, and finally the y intercept, b, by replacing x and y by 2 and 3.

Notice how in both parts (a) and (b) the formulae are copied from the booklet and then substituted with values from the problem, to give the correct answers.

Working:

$$M = \left(\frac{x_1 + x_2}{2}, \frac{y_1 + y_1}{2}\right)$$

$$2 = \frac{s + (-2)}{2}$$

$$s = 4 + 2$$

$$s = 6$$

$$3 = \frac{8 + t}{2}$$

$$t = 6 - 8$$

$$t = -2$$

(b) $m = \dfrac{y_2 - y_1}{x_2 - x_1}$

$= \dfrac{-2 - 8}{-2 - 6}$

$= 1.25$

$3 = -\dfrac{1}{1.25}(2) + b$

$b = 4.6$

Answers:

(a) $s = 6 \qquad t = -2$ **C2**

(b) $y = -\dfrac{1}{1.25}(x) + 4.6$ **C4**

Examiner report

A common mistake students encountered was confusing s and t, and students could not easily find their values. To avoid this error, they had to use the formula for the mid-point and solve a pair of linear equations.

In the second part of the question, most students worked well until they got the gradient of the perpendicular line, but then they got lost in the part where they had to substitute by the coordinates of the mid-point to find the y intercept.

2. *[Maximum mark: 18]*

(i) Jenny has a circular cylinder with a lid. The cylinder has height 39 **cm** and diameter 65 **mm**.

(a) Calculate the volume of the cylinder **in cm³**. Give your answer correct to **two** decimal places. *[3 marks]*

The cylinder is used for storing tennis balls.
Each ball has a **radius** of 3.25 cm.

(b) Calculate how many balls Jenny can fit in the cylinder if it is filled to the top. *[1 mark]*

(c) (i) Jenny fills the cylinder with the number of balls found in part (b) and puts the lid on. Calculate the volume of air inside the cylinder in the spaces between the tennis balls.

(ii) Convert your answer to (c) (i) into cubic metres. *[4 marks]*

(ii) An old tower (BT) leans at 10° away from the vertical (represented by line TG).

The base of the tower is at B so that $M\hat{B}T = 100°$.
Leonardo stands at L on flat ground 120 m away from B in the direction of the lean.
He measures the angle between the ground and the top of the tower T to be $B\hat{L}T = 26.5°$.

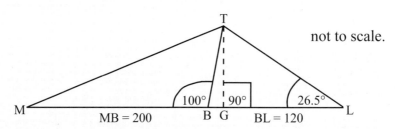

(a) (i) Find the value of angle $B\hat{T}L$.

(ii) Use triangle BTL to calculate the sloping distance BT from the base, B to the top, T of the tower. *[5 marks]*

(b) Calculate the vertical height TG of the top of the tower. *[2 marks]*

(c) Leonardo now walks to point M, a distance 200 m from B on the opposite side of the tower. Calculate the distance from M to the top of the tower at T. *[3 marks]*

[Taken from paper 2 May 2007]

How do I approach the question?

(i) You can make a diagram of a cylinder and put the information given in it. Notice that the units are in **bold**. This stresses their importance. Note that one is given in cm and the other in mm.

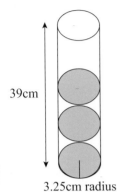

(a) Use the information booklet for this part. You have the formula of the cylinder volume. All the measurements in the formula must have the same unit, and the answer must be given correct to two decimal places.

(b) Again you need another diagram for this part (maybe the diagram done in (a) but now with the tennis balls inside). Look at the mark allocation: this part is worth just 1 mark, meaning that there is not much to do. Only division is required here.

(c) The volume of air is the difference between the volume of the cylinder and the volume occupied by the balls.

(ii) Thoroughly read the question and see if you can link what you are told in words with what you are told in the diagram. You can also see if there is any information that is not written in the diagram, such as angle BGT, which is clearly 90°, or angle BTG, which is 10°. Once you have identified this, you are ready to start answering the question. This is a question on trigonometry and it is very important to identify the angles and triangles that you need to work with. It will help you to draw and label the relevant triangles in your answer sheet and write the information given on them.

What are the key areas from the syllabus?

- Approximation
- Conversion between different units
- Three-dimensional geometry
- Trigonometry

This answer achieved 3/18

Note that the conversion that the student makes from millimetres (mm) to centimetres (cm) is wrong, as 65 mm equals 6.5 cm. Also, to gain the mark for accuracy, the intermediate answer, which in this case would have been $\pi(0.325)^2 \times 39 = 12.9413$, with more than two decimal places, should have been shown.

The unit of 3.25 is centimetres, so why has the student changed it to cm³? The student seems to know that it is not possible to subtract cm from cm³ and that is why the student has changed it. However, 3.25 cm and 3.25 cm³ represent completely different things (3.25 cm represents the radius of the tennis ball and not its volume). This is a good example to appreciate how important the units are.

The student did not calculate the volume of a tennis ball in this part question nor in (b). In (c)(ii) the student converted from cm to m by dividing by 100 but not from cm³ to m³ for which it is necessary to divide by 100 three times.

Within the working, we can see a good way to distinguish between angles BTL and T using different symbols, $\angle$ and $\hat{\,}$. Note that the last line should say $\angle$BTL, but there is no penalty for this.

Triangle BTL is not a right-angled triangle, so you cannot use the tangent here.

Working:

2 (i) (a) $h = 39$ cm

$d = 65$ mm, 0.65 cm

v of a cylinder $= \pi r^2 h$ **M1**

$= \pi(0.325)^2 39$ **A0**

$= 12.94$ cm³   **A0**

(b) 12.94 cm³ $- 3.25$ cm³ $= 9.69$

9.69 cm³ $- 3.25$ cm³ $= 6.44$

6.44 cm³ $- 3.25$ cm³ $= 3.19$

∴ 3 tennis balls will fit in the cylinder   **A0**

(c) (i) 3.19 cm³ is the empty spaces between balls   **M0 A0**

(ii) 0.0319 m³   **A0**

(ii) (a) (i) $B\hat{T}L = 90° + 26.5° + \angle T = 180°$

∴ $\angle T = 180 - 90 - 26.5$

∴ $\angle T = 63.5 + 10\% = 73.5\%$ **M1 A1**

(ii) $\tan 26.5 = \dfrac{BT}{120}$

∴ $BT = 120 \times \tan 26.5$

$BT = 59.83$ **M0 A0**

(b) $\tan 63.5 = \dfrac{120}{TG}$ **M0**

∴ $TG = 59.83$ **A0**

(c) $200^2 + 59.83^2 = MT^2$

∴ $43579.63^2 = MT^2$ **M0 A0**

$208.76 = MT$ **A0**

120 is not the length of GL but the length of BL.

Pythagoras' theorem can be used only in right-angled triangles and MBT is not right-angled. To find the length of MT, the student should have used the cosine rule.

This answer achieved 11/18

If the student had realized that nine balls do not fit into the cylinder, they could have worked out the correct answer by dividing the height of the cylinder by the diameter of the ball and getting the correct answer, which is six balls.

The working is correct, as the student finds the volume occupied by the number of balls found in (b) and takes away that value from the volume of the cylinder. The student follows through from the answer given in (b), so full marks were awarded. In (ii) the conversion made follows through from the previous answer and the student gets the mark as a result of using that $1\,cm^3 = \frac{1}{1000\,000}\,m^3$.

Full marks were awarded because the method followed and the answer found were correct. However, the student was not careful when writing down the calculation for BTL. The correct expression should have been $\angle BTL = 180 - (90 + 26.5) + 10$.

In triangle GTL the length of GL is not 120, so the working here is incorrect.

Working follows through from previous answer. There is correct use of the cosine rule to find the length of MT.

An accuracy penalty is applied for not giving the answer correct to three significant figures.

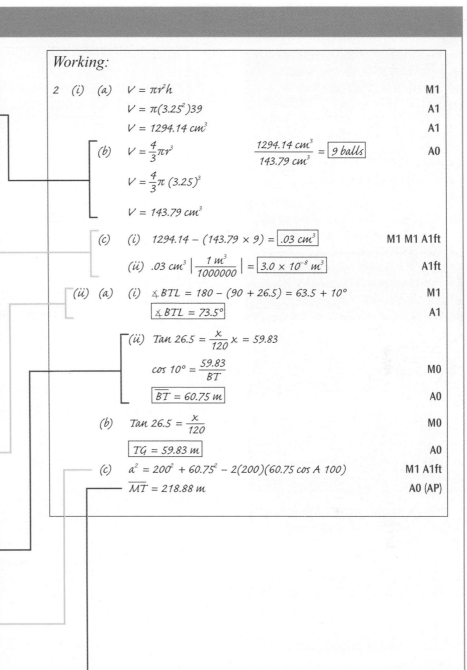

Working:

2 (i) (a) $V = \pi r^2 h$ **M1**
           $V = \pi(3.25^2)39$ **A1**
           $V = 1294.14\,cm^3$ **A1**

(b) $V = \frac{4}{3}\pi r^3$ $\frac{1294.14\,cm^3}{143.79\,cm^3} = \boxed{9\ balls}$ **A0**
     $V = \frac{4}{3}\pi(3.25)^3$
     $V = 143.79\,cm^3$

(c) (i) $1294.14 - (143.79 \times 9) = \boxed{.03\,cm^3}$ **M1 M1 A1ft**

   (ii) $.03\,cm^3 \left|\frac{1\,m^3}{1000000}\right| = \boxed{3.0 \times 10^{-8}\,m^3}$ **A1ft**

(ii) (a) (i) $\angle BTL = 180 - (90 + 26.5) = 63.5 + 10°$ **M1**
           $\boxed{\angle BTL = 73.5°}$ **A1**

    (ii) $\tan 26.5 = \frac{x}{120}$ $x = 59.83$

         $\cos 10° = \frac{59.83}{BT}$ **M0**

         $\boxed{BT = 60.75\,m}$ **A0**

(b) $\tan 26.5 = \frac{x}{120}$ **M0**
    $\boxed{TG = 59.83\,m}$ **A0**

(c) $a^2 = 200^2 + 60.75^2 - 2(200)(60.75\cos A\ 100)$ **M1 A1ft**
    $\overline{MT} = 218.88\,m$ **A0 (AP)**

This answer achieved 17/18

Before starting to answer the question, the student writes down all the information given and shows the working step by step. First of all the student finds the radius of the ball, then substitutes this value into the formula of the volume of the sphere and finally gives the answer correct to two decimal places.

The student loses 1 mark. It was almost a perfect score! This student made the same mistake of calculating the number of balls by using the volume rather than by using the height of the tube. A diagram would have helped the student to understand the situation.

The student showed excellent working and achieved all the follow-through marks for showing all workings.

The student uses the sine rule with correct substitutions and finds the correct answer, so full marks are awarded.

To find the vertical height GT, the student works with the right-angled triangle BGT with angles 90°, 80° and 10°. By using the trigonometric ratio $\sin 80°$, the correct answer is reached.

The student finds the length of MT through the correct use of the cosine rule.

Working:

2 (i) (a) $V_{cyl} = \pi r^2 h$ $h = 39\ cm$ $r = \dfrac{6.5\ cm}{2} = 3.25\ cm$ **M1 A1**

$= \pi \times (3.25^2)(39)$

$\boxed{V_{cyl} = 1294.14\ cm^3}$ **A1**

(b) $V_{sphere} = \dfrac{4}{3}\pi r^3$ $\dfrac{1294.14\ cm^3}{143.79\ cm^3} = \boxed{9\ balls}$ **A0**

$V_{sph} = \dfrac{4}{3}\cdot \pi \cdot (3.25^3)$

$V_{sph} = 143.79\ cm^3$

(c) (i) $9\ balls \cdot \dfrac{143.79\ cm^3}{1\ ball} = 1294.11\ cm^3$ **M1**

$1294.14\ cm^3 - 1294.11\ cm^3 = \boxed{.03\ cm^3}$ **M1 A1ft**

(volume of air in cylinder)

(ii) $.03\ cm^3 = \boxed{3 \times 10^{-8}\ m^3}$ **A1ft**

(ii)

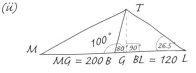

$MG = 200\ B$ G $BL = 120\ L$

(a) (i) $\angle BTL = 180 - (80 + 26.5) = \boxed{73.5°}$ **M1 A1**

(ii)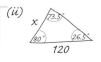

$\dfrac{120}{\sin 73.5} = \dfrac{x}{\sin 26.5}$ $x = 55.8\ m$ **M1 A1 A1**

(b)

$x° = 180 - (80 + 90)$ $x = 10°$ $x + y = 73.5$

$y° = 180 - (26.5 + 90)$ $y = 63.5°$

$\sin 80 = \dfrac{TG}{55.8}$ $\boxed{TG = 55.0\ m}$ **M1 A1**

(c)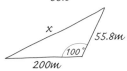

$x^2 = (55.8\ m)^2 + (200\ m)^2 - 2(200\ m)(55.8\ m)(\cos 100)$ **M1 A1**

$x^2 = 46989.47$

$\boxed{x = 217\ m}$ **A1**

Examiner report

For part (ii) you need to select the correct formulae to answer this question. A common error is for students to confuse radius with diameter. Many students divided the cylinder volume by the tennis ball volume to see how many would fit. However, the examiners follow through from this error and award marks where necessary. Students should also be aware of how to convert successfully from cm³ to m³.

A big problem within part (ii) is that students did not carefully read the information given. Many students took GL to be 120, whereas it was clearly written on the diagram that "BL = 120". Also non-right-angled triangles were taken as right-angled triangles. You always need to write down your unit. Having a unit penalty applied for not writing down the units could cost you marks.

9. Statistics

Discrete and continuous data

You should be able to:

- understand the difference between discrete and continuous data.

Example

Discrete data you can count exactly—for example, the number of students in your mathematical studies class, the total score when throwing three dice, and so on.

Continuous data you can measure—for example, height of 13-year-old boys, weight of babies, time taken to run 100 m, and so on.

Frequency tables

You should be able to:

- set up frequency tables and draw frequency polygons for simple discrete data.

Example

The time taken (to the nearest minute) by 30 students to complete a puzzle is given in the table.

Time, t minutes	Frequency, f
1	6
2	10
3	8
4	6

Represent this on a frequency polygon.

Be aware that time is normally continuous. However, the phrase "to the nearest minute" transforms it into discrete.

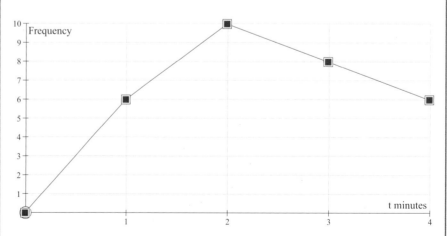

Be prepared

- When plotting a polygon, use the horizontal axis for data and the vertical axis for corresponding frequencies.
- Label your axes and check that the scale on your axes is uniform.

Grouped data, histograms and stem plots

You should be able to:
- set up frequency tables for grouped discrete or continuous data
- draw histograms and stem plots.

Example
Draw a histogram to represent the following information.

Time, t minutes	Frequency
$0 \leq t < 10$	4
$10 \leq t < 20$	7
$20 \leq t < 30$	6
$30 \leq t < 40$	9
$40 \leq t < 50$	4

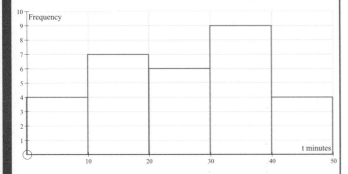

Be prepared
- When drawing a frequency histogram, remember to leave **no** spaces between the bars.
- Check carefully the **key** for a stem-and-leaf diagram. For example, 3|2 could represent 3.2 kg or 32 kg or 32 people, and so on.

Cumulative frequency: box plots

You should be able to:
- set up cumulative frequency tables
- draw a cumulative frequency graph
- find percentiles and quartiles
- draw a box plot.

Example

(a) Draw a cumulative frequency diagram for the previous data.

Time, t minutes	Cumulative frequency
$t < 10$	4
$t < 20$	11
$t < 30$	17
$t < 40$	26
$t < 50$	30

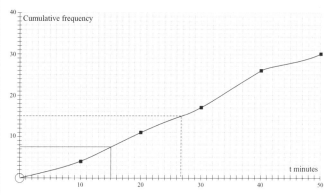

(b) Write down the median and the lower quartile for the data in the previous table.

The total is 30, so, to find the median, you draw a horizontal dotted line through 15 and read where this line cuts the curve. Do the same at 7.5 to find the lower quartile.

Median = 26.8

Lower quartile = 15

Be prepared

- To find the cumulative frequency numbers, you keep adding the frequencies to the ones before. Your last value should be the total frequency—if not, then go back and find your mistake!

- To draw a cumulative frequency curve, always remember to plot the **upper boundary** (not the mid-point) of each group with its corresponding frequency.

- Always indicate the median, quartiles and percentiles on your graph with dotted lines.

- Cumulative frequency curves always go up.

- Join up your points with a smooth curve.

You should know:

- what a **box-and-whisker plot** looks like and what its main features are

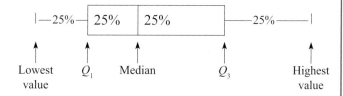

- that **outliers** are points with value either less than $Q_1 - 1.5 \times$ IQR or greater than $Q_1 + 1.5 \times$ IQR, where IQR is the inter-quartile range.

Be prepared

- Draw an axis with a suitable scale for your box-and-whisker plot.

Measures of central tendency

You should know:

- $x_1, x_2, \ldots$ represent different values of x
- $f_1, f_2, \ldots$ are the frequencies corresponding to $x_1, x_2, \ldots$
- $\sum$ means "the sum of"
- $\bar{x}$ indicates the mean
- **mean** for single discrete data is $\bar{x} = \dfrac{\sum_{r=1}^{n} x_r}{n}$, which means add up all the numbers and divide by how many numbers you are given
- **mean** for data with frequencies is $\bar{x} = \dfrac{\sum_{r=1}^{n} f_r x_r}{\sum_{r=1}^{n} f_r}$, which means multiply the numbers by their frequencies, add these numbers up and divide by the total frequency
- how to define mode, modal group and median.

Example

(a) Find the mode, median and mean for the following numbers:

6 8 3 2 1 6 4 9 7 6

First write them in order:

1 2 3 4 6 6 6 7 8 9

There are three sixes, so mode = 6

Median $= \dfrac{(6 + 6)}{2} = 6$

Mean $= \dfrac{(1 + 2 + 3 + 4 + 6 + 6 + 6 + 7 + 8 + 9)}{10} = 5.2$

(b) Find the estimate of the mean for the following:

x	f	m	$m \times f$
$0 \leq x < 10$	2	5	10
$10 \leq x < 20$	3	15	45
$20 \leq x < 30$	5	25	125
$30 \leq x < 40$	3	35	105
$40 \leq x < 50$	2	45	90

Here m is the mid-interval value, which is found by adding the end points of each group and dividing by 2.

Total frequency = 15

Total $m \times f$ = 375

Estimate of mean $= \dfrac{375}{15} = 25$

Be prepared

- The mid-interval value is found by adding the higher and lower boundary values of each interval and dividing the answer by 2.

Measures of dispersion

You should know:

- range = largest value − smallest value
- inter-quartile range = IQR = $Q_3 - Q_1$
- standard deviation = $s_x = \sqrt{\dfrac{\sum_{i=1}^{k} f_i(x_i - \bar{x})^2}{n}}$, where $n = \sum_{i=1}^{k} f_i$

Example

(a) Write down the standard deviation of the following numbers:

3   5   8   6   2   9   1   7   3

s.d. = 2.64

Texas Instruments

STAT – CALC – 1-var stats L_1

This gives you mean, standard deviation, median, and so on for a **single** list of data.

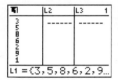
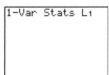

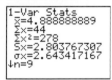

Casio

STAT – CALC – 1VAR

(SET – make sure that XList is List 1 and Freq is 1)

This gives you mean, standard deviation, median, and so on for a **single** list of data.

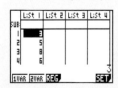

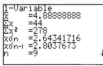

(b) Write down the standard deviation of the following numbers x, where f is the frequency of occurrence:

x	3	6	9	12
f	2	8	5	3

s.d. = 2.69

Texas Instruments

STAT – CALC – 1-var stats L_1, L_2

This gives mean, and so on for a list of data and corresponding frequencies.

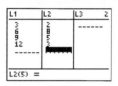
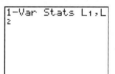

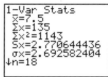

Casio

STAT – CALC – SET – XList: List 1; Freq: List 2 – EXE – 1VAR

This gives mean, and so on for a list of data and corresponding frequencies.

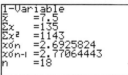

Be prepared

- Different calculators have different symbols for standard deviation and there are also two different standard deviations given. Remember that you should use the **smaller** one in your exam papers.

9. Statistics

Scatter diagrams and correlation coefficient

You should be able to:
- draw scatter diagrams and the line of best fit by eye
- comment on the result of the correlation coefficient, r.

You should know:
- in most questions you will be able to find r straight from your GDC
- you will always be given s_{xy} and the values for s_x and s_y you can find from your GDC.

Example
Write down the correlation coefficient for the following data and comment on the result.

Weight	63	54	46	71	89	42	68	55
Height	160	152	146	150	173	150	168	155

Texas Instruments

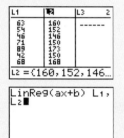

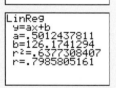

Casio

CALC – REG –

r = 0.799 so there is a strong positive correlation.

Be prepared
- The line of best fit by eye must pass through the mean point. It does not usually pass through the origin.
- The value of r must lie between −1 and +1, where −1 is a perfect negative correlation and +1 is a perfect positive correlation.
- Use the proper mathematical terminology when describing r. The value for r can either imply no correlation or imply weak, moderate, strong, very strong, perfect, positive or negative correlation.

Regression line

You should know:
- equation of **regression line** of y on x is
$$(y - \bar{y}) = \frac{s_{xy}}{s_x^2}(x - \bar{x})$$

Example
Write down the regression line for the data in the previous example.

Using the same windows as for the correlation coefficient we get
y = 0.501x + 126 (see the GDC screenshots in the previous example).

Be prepared
- Only use your regression line for prediction purposes if there is at least a moderately strong correlation.
- Do not predict values outside the given data range.

9. Statistics

Chi-squared test

You should be able to:
- calculate and use the χ^2 test
- calculate the degrees of freedom and the expected values.

You should know:
- H_0, the null hypothesis, states that the two variables are independent and H_1, the alternative hypothesis, states that they are not independent
- if the χ^2_{calc} is less than the critical value then you accept H_0
- on your GDC you will also be given the *p*-value—if the *p*-value is more than the significance level then you accept H_0
- $\chi^2_{calc} = \sum \dfrac{(f_o - f_e)^2}{f_e}$
- degrees of freedom = (number of rows − 1) × (number of columns − 1)
- expected value = $\dfrac{\text{(row total)} \times \text{(column total)}}{\text{overall total}}$
- the critical values of the χ^2 distribution—for this you can use your information booklet.

Example

The following information was collected and shows the type of chocolate bar preferred by 40 men and 60 women. By using a chi-squared test at the 5% significance level, determine if chocolate bar preference and gender are related.

	Plain	Nut	Filled	Totals
Male	10	14	16	40
Female	22	30	8	60
Totals	32	44	24	100

H_0: Chocolate bar preference and gender are independent

H_1: Chocolate bar preference and gender are not independent

Degrees of freedom = $(2-1) \times (3-1) = 2$

Expected values: $\dfrac{\text{(row total)} \times \text{(column total)}}{\text{overall total}}$

For example, for "plain" and "male" it is $\dfrac{40 \times 32}{100} = 12.8$

	Plain	Nut	Filled	Totals
Male	12.8	17.6	9.6	40
Female	19.2	26.4	14.4	60
Totals	32	44	24	100

Chi-squared value = $\dfrac{(10 - 12.8)^2}{12.8} + \ldots + \ldots$

= 9.36

Texas Instruments

MATRIX
Use this to put in your observed values for the χ^2 test.

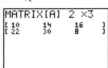

STAT – TESTS – χ^2-test

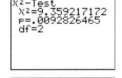

Casio

MENU – RUN-MAT – MAT – DIM – m:3; n:3 – EXE

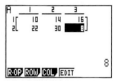

MENU – STAT – TEST – CHI

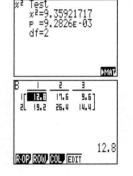

Critical value = 5.991

9.36 > 5.991 so we do not accept the null hypothesis.

Or you could use the fact that the p-value, 0.009 28 < 0.05, so we do not accept the null hypothesis.

This means that chocolate bar preference and gender are not independent.

Be prepared

- Find (from the table in your information booklet) the **critical value** for the given **significance level**. This is usually 5%, but it can also be 1% or 10%.
- If the expected values are less than 5, then you have to combine rows or columns. Remember that this will affect your table of observed values and your degrees of freedom.

2. (a) State which of the following sets of data are discrete.

   (i)   Speeds of cars travelling along a road.
   (ii) Numbers of members in families.
   (iii) Maximum daily temperatures.
   (iv) Heights of people in a class measured to the nearest cm.
   (v)   Daily intake of protein by members of a sporting team.

   The boxplot below shows the statistics for a set of data.

   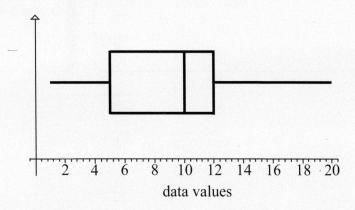

   data values

   (b) For this data set write down the value of

   (i)   the median

   (ii) the upper quartile

   (iii) the minimum value present

   (c) Write down three different integers whose mean is 10.

   [Taken from paper 1 May 2007]

9. Statistics

How do I approach the question?

This is a question from paper 1.

(a) You are required to deal with discrete and continuous data and to understand the difference between the two. Read each of the sets carefully before deciding which ones are discrete.

(b) You need to remember what values the vertical lines of the box plot represent and also that the ends of the whiskers represent the minimum and maximum entries, respectively.

(c) You must think of three different integers (positive or negative) that add up to 30.

What are the key words within this question?

In order to help with this question, it is important to remember "to the nearest cm". This will help you to avoid losing marks.

What are the key areas from the syllabus?

- Box plot
- Measures of central tendency and dispersion

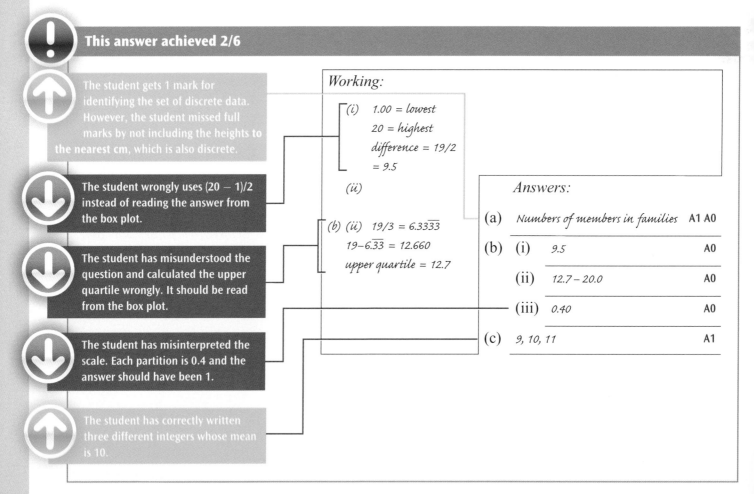

This answer achieved 2/6

The student gets 1 mark for identifying the set of discrete data. However, the student missed full marks by not including the heights to the nearest cm, which is also discrete.

The student wrongly uses (20 − 1)/2 instead of reading the answer from the box plot.

The student has misunderstood the question and calculated the upper quartile wrongly. It should be read from the box plot.

The student has misinterpreted the scale. Each partition is 0.4 and the answer should have been 1.

The student has correctly written three different integers whose mean is 10.

Working:

(i) 1.00 = lowest
20 = highest
difference = 19/2
= 9.5

(ii)

(b) (ii) 19/3 = 6.3333
19−6.33 = 12.660
upper quartile = 12.7

Answers:

(a) Numbers of members in families — A1 A0

(b) (i) 9.5 — A0

(ii) 12.7 – 20.0 — A0

(iii) 0.40 — A0

(c) 9, 10, 11 — A1

88

This answer achieved 4/6

Daily temperatures (iii) is wrong because these are continuous and the student also missed out (ii) which is discrete.

Although the working out is incorrect, the correct answer is given and awarded 1 mark. The median should have been read from the box plot. The vertical line inside the box plot represents the median, which is 10. The student has taken the maximum value of 20 and divided this by 2. This also resulted in the answer 10, but the method is not correct. The student was very lucky, and was awarded 1 mark because the correct answer was written in the answer space.

The correct values are taken from the box plot and no working out was needed.

The total of the three integers is 30, but they are not **different**, and so no marks are awarded.

Working:

(b) (i) $\dfrac{20}{2} = 10$

(ii) $Q_3 = 12$

(c) $\dfrac{10 + 10 + 10}{3} = 10$

Answers:

(a) *iii, iv*   A0 A1

(b) (i) 10

(ii) 12   C3

(iii) 1

(c) 10, 10, 10   A0

This answer achieved 6/6

Working:

(ii) 2/5 = .4 + .4 + .2 = 1

(c) 30/3 = 10
    6 + 10 + 14 = 30/3

Answers:

(a) ii, iv   C2

(b) (i)   10

    (ii) 12   C3

    (iii) 1

(c) 6, 10, 14   C1

Notice how the student has correctly worked out the scale by dividing 2 by 5 to calculate what each partition is worth.

6, 10 and 14 are three different integers whose mean is 10.

The student correctly identified both answers as discrete.

Values are read correctly from the box plot.

Examiner report

For part (a) it is important to read the question carefully and pick out key words or terms. Do not skim over it or you may miss out important details like "to the nearest cm", which transforms continuous data into discrete data. Try to remember that discrete data can be counted and continuous data can be measured.

In part (b), you need to be able to read the scale accurately. Each 2 units are divided into 5 parts, so 2/5 = 0.4 for each partition. Remember also that the line inside the box is the median and not the mean.

There are many ways to get the answer to this part—the sum of many three integers add up to 30. However, a common mistake here was putting 10 + 10 + 10. These are not **different** integers.

14. Tania wishes to see whether there is any correlation between a person's age and the number of objects on a tray which could be remembered after looking at them for a certain time.

    She obtains the following table of results.

Age (x years)	15	21	36	40	44	55
Number of objects remembered (y)	17	20	15	16	17	12

    (a) Use your graphic display calculator to find the equation of the regression line of y on x. *[2 marks]*

    (b) Use your equation to estimate the number of objects remembered by a person aged 28 years. *[1 mark]*

    (c) Use your graphic display calculator to find the correlation coefficient r. *[1 mark]*

    (d) Comment on your value for r. *[2 marks]*

    [Taken from paper 1 May 2007]

How do I approach the question?

Be aware that this question is near the end of paper 1. Questions that are at the end of this paper tend to be harder than those at the beginning.

(b) You have to substitute 28 for x in the equation you have found in part (a) and make sure to give your answer as a whole number.

(d) You must use correct mathematical terminology to comment on r, such as positive, negative, weak, moderate or strong.

Which GDC functions will I need?

For part (a) you are told to use your GDC to find the equation of the regression line. You have to remember to put "Age" in one list and "Number of objects remembered" in a second list, and to use both lists to find the regression line.

Texas Instruments

STAT – CALC – LinReg(ax+b) L_1, L_2

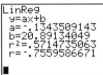

Casio

MENU – STAT – CALC – SET – Xlist: List 1; YList: List 2 – REG-X

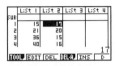

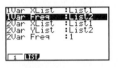

For part (c) you are being told to use your GDC to find the value for r. You should have found this value already in part (a)—see the GDC screens above. If you are using a TI calculator and the value for r did not appear, then make sure that Diagnostic is on. This can be done via "Catalog".

Texas Instruments

This answer achieved 2/6

The working out showed that the student found the correct answer to begin with and then crossed it out and gave another (incorrect) answer. Examiners do not mark crossed-out work.

The working out is correct but the answer is impossible, as you cannot have a negative number of objects. In addition, the answer is not a whole number.

The student has forgotten to put in the negative sign.

This is correct follow through from the student's previous answer and so is awarded full marks.

Working:

(a) ~~$y = ax + b$~~
~~$a = -.34$~~
~~$b = -20.89$~~

~~$y = 134x + 20.89$~~

(a) $a = -4.25$
$b = 103.93$
$\therefore y = -4.25x + 103.93$
$\downarrow$
104 (3 S.F.)

(b) $y = -4.25(28) + 103.93$
$= -199 + 103.93$
$= -15.07$

Answers:

(a) $y = -4.25x + 104$ **A0 A0**

(b) -15.07 **A0**

(c) $.756$ **A0**

(d) r is quite close to 1 $\therefore$ a positive med-strong relationship is evident.   **C2ft**

This answer achieved 4/6

Working out is correct here but the answer is not a whole number and so the student has lost 1 mark.

No marks have been deducted for not having the answer to three significant figures. This means that the student has already lost this penalty mark in an earlier question.

Again, no marks have been deducted for incorrect rounding. The accuracy penalty is only taken off once in the paper.

The student correctly put down negative correlation, but forgot to say that it was also moderately strong.

Working:

(a) $y = ax + b$
$y = -0.134x + 20.891$

(b) $y = -0.134(28) + 20.891$
$= 17.139$

Answers:

(a)	$y = -0.134x + 20.891$	C2
(b)	17.139	A0
(c)	$r = -0.755$	A1
(d)	Negative correlation	A1

This answer achieved 6/6

The answer has been correctly rounded to three significant figures and the student has received full marks.

The correct working out is shown in the box and the answer has been rounded to the nearest whole number. This student realized that the answer had to be a whole number of objects.

The full GDC answer is given in the box and it has been correctly rounded in the answer space.

Correct terminology is used and both answers are given.

Working:

(a) $y = ax + b$
 $a = -0.134350914$
 $b = 20.89134049$
 $y = -0.134x + 20.9$

(b) $y = -0.134 \times 28 + 20.9$
 $y = 17.148$

(c) $r = -0.755958667$

(d) Moderately strong negative correlation

Answers:

(a) $y = -0.134x + 20.9$ — C2

(b) 17 objects — C1

(c) $r = -0.756$ (3 s.f.) — C1

(d) Moderately strong negative correlation. — C2

Examiner report

This question dealt with the correlation coefficient, the regression line and its use for prediction purposes. This type of question highlights common mistakes. Students may be careless when entering the numbers or may not know how to use their GDC correctly. Also students may not write their answer as a whole number. You cannot really have 17.139 objects, can you? Within part (d) marks can easily be lost when students use the wrong terminology, such as fine, good and reasonable. Correct terminology is words like strong, moderate, weak, positive and negative.

4. *[Maximum mark: 20]*

(i) A random sample of 167 people who own mobile phones was used to collect data on the amount of time they spent per day using their phones. The results are displayed in the table below.

Time spent per day (t minutes)	$0 \leq t < 15$	$15 \leq t < 30$	$30 \leq t < 45$	$45 \leq t < 60$	$60 \leq t < 75$	$75 \leq t < 90$
Number of people	21	32	35	41	27	11

(a) State the modal group. *[1 mark]*

(b) Use your graphic display calculator to calculate approximate values of the mean and standard deviation of the time spent per day on these mobile phones. *[3 marks]*

(c) On graph paper, draw a fully labelled histogram to represent the data. *[4 marks]*

(ii) Manuel conducts a survey on a random sample of 751 people to see which television programme type they watch most from the following: Drama, Comedy, Film, News. The results are as follows.

	Drama	Comedy	Film	News
Males under 25	22	65	90	35
Males 25 and over	36	54	67	17
Females under 25	22	59	82	15
Females 25 and over	64	39	38	46

Manuel decides to ignore the ages and to test at the 5 % level of significance whether the most watched programme type is independent of **gender**.

(a) Draw a table with 2 rows and 4 columns of data so that Manuel can perform a chi-squared test. *[3 marks]*

(b) State Manuel's null hypothesis and alternative hypothesis. *[1 mark]*

(c) Find the expected frequency for the number of females who had 'Comedy' as their most-watched programme type. Give your answer to the nearest whole number. *[2 marks]*

(d) Using your graphic display calculator, or otherwise, find the chi-squared statistic for Manuel's data. *[3 marks]*

(e) (i) State the number of degrees of freedom available for this calculation.

(ii) State the critical value for Manuel's test.

(iii) State his conclusion.

[3 marks]

[Taken from paper 2 November 2007]

How should I approach the question?

This question is divided into two parts: part (i) tests your knowledge of measures of central tendency and dispersion plus histograms, while part (ii) is on the chi-squared test.

(i) (a) You need to write down the group with the highest frequency.

(b) You have to remember how to use your GDC to find the approximate mean and standard deviation from a grouped frequency table.

Texas Instruments

STAT – CALC – 1-var stats L_1, L_2

This gives mean, and so on for a list of data and corresponding frequencies.

Casio

STAT – CALC – SET – XList: List 1; Freq: List 2 – EXE – 1VAR

This gives mean, and so on for a list of data and corresponding frequencies.

(c) When creating histograms, check that you are using the correct scales on your axes.

(ii) (a) You are asked to draw another table with two rows and four columns. (Do not mix up rows and columns—you sit in a row at the cinema, and a column is a vertical construction). But this time collect the data for the males together and the females together.

Texas Instruments

Casio

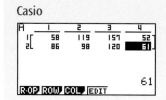

(b) You need to remember that the null hypothesis states that the two variables are independent and that the alternative hypothesis states that they are not independent.

(c) To find this expected value, multiply the total number of Females by the total of Comedy, and divide by 751.

(d) You are told to use your GDC to find the chi-squared statistic.

Texas Instruments

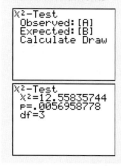

Casio

(e) (i) To find the number of degrees of freedom, multiply (rows – 1) by (columns – 1).

(ii) Use your **information booklet** and remember to look under the correct column (0.95 in this case) and the correct row, which is given by the number of degrees of freedom (here 3).

(iii) Do not forget to give a reason for your conclusion. Your conclusion will either be that you accept the null hypothesis or that you do not.

What are the key words within this question?

- "Approximate"—When you enter the data for the time into list 1, you have to put in the mid-point of each group. Not all people in the group will speak for that amount of time. Some will be lower and some will be higher. So the value you get for the mean and standard deviation will only be approximate and not exact.

Which GDC functions will I need?

This question is a prime example for you to use your GDC skills and knowledge.

- Calculate mean and standard deviation.
- Know how to find the chi-squared test statistic.
- The degrees of freedom is also given on the GDC. If you are asked to show it, then it is (number of rows − 1) × (number of columns − 1).
- Make sure that you know how to calculate the expected values and to give your answer to the nearest whole number. You are sometimes asked to show how to find one of these values. Of course, you do not have to work them all out by hand, as your GDC does this for you.

Texas Instruments

MODE − FLOAT − 0 will give your answer to the nearest integer. But remember to change it back again for the next question:

Casio

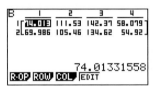

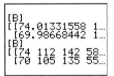

This answer achieved 8/20

Working:

4 (i) (a) modal group $45 \leq t < 60$ A1

(b) mean values of t minutes = <u>45</u>   G0

Mean values of number of people = 27.8   G0

std deviation of time = <u>25.6</u>   G0

(a) (ii)

	Drama	Comedy	Film	News
Males	58	119	157	52
Females	86	98	120	61

M2
A1

(b) null hypothesis = the most watched programme is independent of gender   A1

alternative = the most watched programme depends on the gender

(ii) (c) <u>130</u>   G0

(d) Chi-squared = 68.52   G0

(e) (i) df = 3   A1

(ii) Critical value = 0.072   A0

(iii) If X^2 calc < critical value accept Ho therefore Manuels rejects Ho and accepts alternative hypothesis   A1ft

(i) (c)

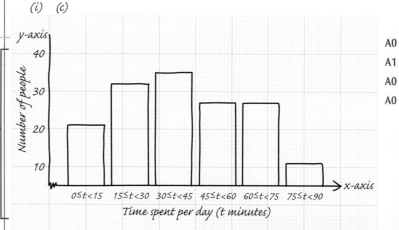

A0
A1
A0
A0

Comments:

- It is relatively easy to find out the modal group. Ensure that you are familiar with the term and what is required.

- The student should have put the mid-point of the times into List 1 and the frequencies into List 2. The answer is wrong, so the student has not used the GDC correctly.

- The new table was clearly drawn with two rows and four columns and correct data for full marks.

- The student has correctly written down the null and alternative hypotheses.

- The final conclusion follows through from the numbers that the student has found and so is awarded the mark.

- This is read incorrectly from the column, 0.005 instead of 0.95.

- The student really struggled with this graph. The scales on **both** axes are wrong, there are spaces between the bars, and one of the bars is drawn incorrectly.

- This value is wrong and there is no working, so no method marks can be awarded. Try to show some working, as it may be right and you will get 1 mark. The correct answer is 105.

- This value is nowhere near the correct answer. The student should have put the observed values into a 2 by 4 matrix and then used the chi-squared test on their GDC.

- The student did get 1 mark for putting the correct labels on the axes.

- The student wrote down the correct answer for the degrees of freedom.

9. Statistics

This answer achieved 11/20

Regardless of which GDC the student used, these are both wrong. The student put the **end values** into List 1 when they should have put the **mid-values** into it, and the number of people into List 2.

This answer highlights that the student does not know what the null and alternative hypotheses are.

The student has shown working out for the answer, but both are incorrect. The student has totalled the Female column, divided the number of Females who like Film by it, and then multiplied by 100. This was not what was asked for.

We have no idea where this value came from—but it is not correct!

The student is mixing up the correlation coefficient with the chi-squared test statistic and gets no marks for this answer.

The graph has been neatly drawn with a ruler and the axes are correctly labelled, which is why the student was awarded 3 marks.

The scale on the horizontal axis is not correct and the student has lost 1 mark. You do not write $0 \leq t < 15$ under the bar. Instead, you put the numbers in the same way as on the vertical axis.

Working:

4   (i)   (a) $45 \leq t < 60$ A1

        (b)   mean = 52.5   G0

            standard deviation = 25.6   G0

(ii) (a)

	Drama	Comedy	Film	News	
MALE	58	119	157	52	M2
Female	86	98	120	61	A1

(b) $0.00569 \rightarrow 0.00570$ A0

(c) $86 + 98 + 120 + 61 = 364$ $\dfrac{120}{364} \times 100$

                                       = 32.96

                                       = 33%   A0

(d) $X^2 = 12.6$ G3

(e)   (i)   3   A1

      (ii)   0.0057   A0

      (iii) The correlation is weak.   A0

A0
A3

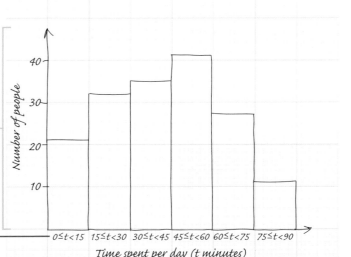

This answer achieved 19/20

> The student has found the mid-points of the groups and shown working for the first one of these. Both answers are correct, but the student has missed out the units, and so loses 1 mark here. Otherwise it would have been a perfect score!

> No working out has been shown here, but the answer is correct from the GDC and it has been given to the nearest whole number, so full marks are awarded. The student has written down the full table of expected values from their GDC.

> The student has correctly used the GDC to find the chi-squared test statistic.

> The student has given a good reason for rejecting the null hypothesis.

> A perfect way to present a graph. It is neatly drawn with a ruler and the axes are correctly labelled, with the correct scales on each axis.

Working:

4 (i) (a) $45 \leq t < 60$ **A1**

(b) $15/2 = 7.5$
$\bar{x} = 42.4$ **G1 G0**
$6x = 21.6$ **G1**

(ii) (a)

observed:

	Feo	C	F	N
M	58	119	157	52
F	86	98	120	61

Expected:

	Feo	C	F	N
M	74	112	142	58
F	70	105	135	55

M2
A1

(b) Ho type of tv most watched is independent of gendre.
H1 the tv watched is not independent of gendre.   **A1**

(c) F & C = 105   **G2**

(d) $X^2 = 12.6$ **G3**

(e) (i) $df = (4-1)(2-1)$
$df = 3 \times 1 = 3$ **A1**

(ii) X^2 crit $= 7.815$ **A1**

(iii) X^2 calc $> X^2$ crit & reject Ho   **A1**

A4

(Histogram: Number of people vs Time spent per day (mins))

Examiner report

This question related to students' understanding of what was meant by a modal group and how efficiently they could use their GDC to find the approximate mean and standard deviation from grouped data. A common problem is that students do not know how to use the mid-point of each group as their List 1 data, and these answers indicate that some students do not know how to use their GDC properly. You can see how important understanding how to use your GDC can be.

Many students calculated the mean by hand, sometimes accurately and sometimes not, and they also tried to find the standard deviation by hand, which can lead to wasted time—even though the question told them to use their GDC.

A common problem with histograms, where marks are easily lost, is incorrect labelling of scales. The bars themselves were usually correct, with the exception of those students who left spaces between the bars and those who drew line graphs instead.

Three relatively easy marks were offered for creating a table combining the males and females, but alternatively some marks were lost for students who put the data into a 4 by 2 instead of a 2 by 4 layout. Take your time reading the seemingly "easier" questions.

An understanding of the chi-squared test was required for the last part. Students who knew what the chi-squared test was managed to write down the null and alternative hypotheses correctly. Many students could also give the expected frequency but not all of them wrote their answer to the nearest whole number and so lost the mark.

When accepting or rejecting the chi-squared test, you can either use the fact that the chi-squared value is less than or greater than the critical value or, using the p-value from the GDC, p is either greater than or less than the significance level (in this case 5% or 0.05).

This type of question lets you show the examiner your full range of skills, efficiency in using your GDC and presenting graphs.

10. Introductory differential calculus

Finding the gradient of a straight line

You should be able to:

- find the gradient of a line joining two points (chord) on a curve.

You should know:

- the gradient of the line joining $P(x_1, y_1)$ and $Q(x_2, y_2)$ is
$$m = \frac{y_2 - y_1}{x_2 - x_1}$$

Example

Find the gradient of the line joining the points $A(-3, 15)$ and $B(4, 5)$.

$$m_{AB} = \frac{5 - 15}{4 - (-3)} = \frac{-10}{7}$$

The derivative as the gradient function

You should be able to:

- find the gradient of a curve at a specific point.

You should know:

- the derivative is the gradient of the chord joining two points on a curve, as the second point is brought close to the first point. It is found from *first principles* using the rule

$$f'(x) = \lim_{h \to 0} \left(\frac{f(x+h) - f(x)}{h} \right)$$

[Note that differentiation from first principles will not be assessed in exams, so there is no need to panic!]

- the gradient of the tangent at a point P is the same as the gradient of the curve at point P

- if using $f(x)$ notation, the derivative is referred to as $f'(x)$

- if using y notation, the derivative is referred to as $\frac{dy}{dx}$.

103

Finding the derivative

You should be able to:

- find the derivative of a polynomial function of one or several terms using the rule
- find the derivative of terms with both positive and negative integer values as the exponent.

You should know:

- $f'(x), \frac{dy}{dx}, \frac{dx}{dt}$, and so on, are used to represent the derivative function
- if $f(x) = ax^n$, then $f'(x) = anx^{n-1}$ and $f''(x) = an(n-1)x^{n-2}$
- the derivative of a constant term is zero.

Example

(a) Find $f'(x)$ if $f(x) = \frac{x^3 + 3x^2 - 6}{3}$.

To answer this question, you must first divide each of the terms in the numerator by 3. This gives a much easier function to differentiate.

$f(x) = \frac{1}{3}x^3 + x^2 - 2$

$f'(x) = x^2 + 2x$

(b) Hence find $f'(-2)$.

$f'(-2) = (-2)^2 + 2 \times (-2)$

$f'(-2) = 4 - 4$

$f'(-2) = 0$

When you are calculating $f'(-2)$, you are finding the value of the gradient at the point $x = -2$ on the curve.

Be prepared

- If x is in the denominator, rewrite it as a negative exponent, then differentiate the term, for example

$f(x) = \frac{3}{x^4} = 3x^{-4}$

- If $f(x)$ is given as a quotient, first divide and then differentiate term by term, for example

$f(x) = \frac{2x^2 - 3x - 4}{2} = x^2 - \frac{3}{2}x - 2$

$f'(x) = 2x - \frac{3}{2}$

Gradients of curves

You should be able to:
- find the gradient of a curve at a given point
- determine the point on a curve corresponding to a particular gradient
- find the equation of the tangent at a given point on a curve.

You should know:
- that parallel lines have the same gradient.

Example

If $f(x) = -x^2 - x$, find the equation of the tangent at $x = 4$.

The gradient is

$f'(x) = -2x - 1$

$f'(4) = -2 \times 4 - 1 = -8 - 1 = -9$

At $x = 4$: $y = -(4)^2 - 4 = -20$

$y = -9x + c$

$-20 = -9 \times (4) + c$

$c = 16$

so the equation of the tangent at $x = 4$ is

$y = -9x + 16$

Texas Instruments	Casio
DRAW – Tangent – 4 – Enter	SKETCH – Tang – 4 – EXE – EXE (make sure that Derivative is ON in the SET UP menu)

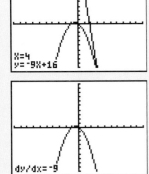

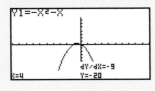

Increasing and decreasing functions

You should be able to:
- recognize when the graph of a function is increasing, decreasing or stationary
- determine from the derivative function the values of x for which the function is increasing or decreasing
- solve real-world problems involving increasing and decreasing functions.

You should know:
- that a function is increasing if $f'(x) > 0$, decreasing if $f'(x) < 0$ and stationary if $f'(x) = 0$.

Example

A function is given by the equation $f : x \mapsto -2x^3 - 2x + 1$. Determine if the function is increasing or decreasing at $x = 2$.

$f'(x) = -6x^2 - 2$

$f'(2) = -6 \times 2^2 - 2 = -26$

$-26 < 0$, hence the function is decreasing at $x = 2$.

Local maximum and minimum points

You should be able to:

- find the values on a curve where the gradient is zero
- find the value of the function at the maximum or minimum point
- solve real-world problems involving maxima and minima points.

You should know:

- that a function has a local maximum or minimum point (stationary point) when $f'(x) = 0$.

Example

A farmer wishes to enclose a rectangular field using an existing fence as one of the four sides.

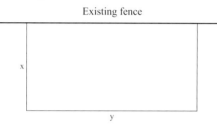

(a) Write an expression in terms of x and y that shows the total length of the new fence. Lengths x and y are measured in metres.

$2x + y$

(b) The farmer has enough materials for 250 m of new fence. Show that $y = 250 - 2x$.

$250 = 2x + y$

$250 - 2x = y$

(c) $A(x)$ represents the area of the field in terms of x.
   (i) Show that $A(x) = 250x - 2x^2$.

   Area $A(x) = xy = x(250 - 2x) = 250x - 2x^2$

   (ii) Find $A'(x)$.

   $A'(x) = 250 - 4x$

   (iii) Hence or otherwise find the value of x that produces the maximum area of the field.

   At maximum point $A'(x) = 0$.

   $0 = 250 - 4x$

   $4x = 250$

   $x = 62.5\,m$

(iv) Find the maximum area of the field.

$A(x) = 250x - 2x^2$

$A(62.5) = 250 \times 62.5 - 2 \times (62.5)^2$

$= 7812.50$

$= 7810\,m^2$

Texas Instruments

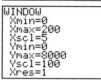

CALC – Maximum

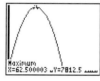

Casio

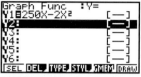

G-Solve – MAX – EXE

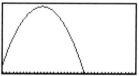

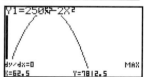

Continued

Local maximum and minimum points (continued)

Example

To find the value of $\frac{dy}{dx}$ for given values of x. For example, if $y = 3x^2 - 8x + 1$, find the value of $\frac{dy}{dx}$ when $x = 2$.

Texas Instruments	Casio
Sketch in standard window	Draw in standard window
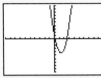	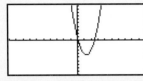
CALC – dy/dx – 2	Sketch – Tang – 2 – EXE
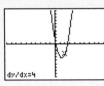	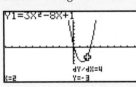

To find the equation of a tangent. For example, find the equation of the tangent to the curve $y = x^2$ at the point where $x = 2$.

Texas Instruments	Casio
2nd PRGM (DRAW) 5: Tangent	SKETCH – Tang – 2 – EXE – EXE (make sure that Derivative is ON in the SET UP menu)
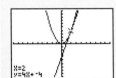	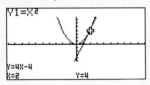

11. The figure below shows the graphs of functions $f_1(x) = x$ and $f_2(x) = 5 - x^2$.

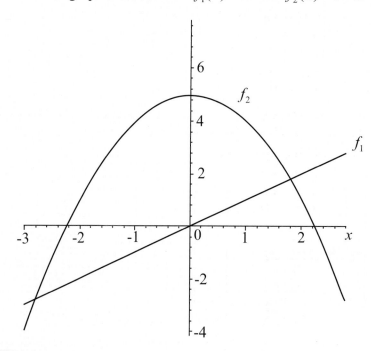

(a) (i) Differentiate $f_1(x)$ with respect to x.

   (ii) Differentiate $f_2(x)$ with respect to x.

(b) Calculate the value of x for which the gradient of the two graphs is the same.

(c) Draw the tangent to the **curved** graph for this value of x on the figure, showing clearly the property in part (b).

[Taken from paper 1 May 2007]

How do I approach the question?

A lot of information is provided in the equations and on the graph in this question. On the graph, f_1 is associated with the equation $f_1(x) = x$ and the parabola is the graph of $f_2(x) = 5 - x^2$.

(a) This can be answered without reference to the graphs of the functions, but the graphs are useful in answering parts (b) and (c).

(c) The question is stressing that you are to draw the tangent to the **curved** graph and not the linear one. This is why the word curved is written in bold.

Usually answers are written in the spaces after the working box. In this question, however, there is only room for answers to parts (a)(i) and (ii) and part (b). Many students missed part (c) because there was not a space in the answer area for it. Part (c) had to be drawn directly onto the graph. Always check that you have answered every part of a question, in the appropriate place on the question booklet.

What are the key areas from the syllabus?

- Differentiation rules
- Finding the equation of a straight line
- Finding the equation for the gradient of a tangent to a curve

This answer achieved 0/6

The student has not differentiated the two functions but simply rewritten them in the answer lines. There is no working to show how the answer to part (b) was obtained. An incorrect answer with no working receives no marks.

The student has not used a ruler to mark in the tangent on the graph in part (c). Tangents are always straight lines. Although it appears that the tangent is drawn at $x = -1$, the follow-through answer to part (b), the tangent is not drawn accurately enough to achieve a mark. The student's tangent is not parallel to f_1 and this should have alerted the student to the possibility of an incorrect answer to part (b).

Working:

$x = 5 - x^2$

Answers:

(a) (i) x A0

(ii) $5 - x^2$ A0

(b) -1 A0

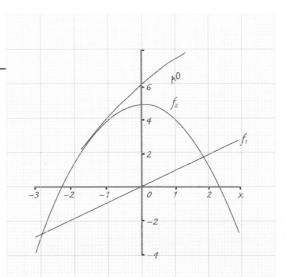

A0

10. Introductory differential calculus

This answer achieved 3/6

In the first part of the question the student has written $f(x) = x^1$ to make finding the derivative easier, using the rule.

The student has made an error in the working box but fortunately has written down the correct answer in the answer lines. In paper 1, if the correct answer is seen in the answer lines, full marks will be awarded regardless of correct or incorrect working. Ideally, both the working in the box and answers should be correct. This student has seen that they have made a mistake and corrected their answer before writing it in the answer lines. If no answer had been given in the answer lines, the examiner would have marked the working and answer in the box. The student would not have achieved full marks in this case.

The student has left part (b) blank and has not drawn any tangent for part (c) and therefore has not received any marks.

Working:

(a) (i) $f(x) = x^1$
         $f_1(x) = 1$

     (ii) $5 - x^2$
         $f_2(x) = 2x$

Answers:

(a)   (i) $f_1(x) = 1$ **A1**

     (ii) $f_2(x) = -2x$ **A1 A1**

(b) _____

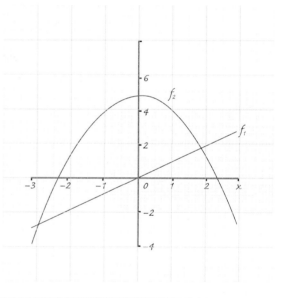

A0

This answer achieved 6/6

↑ The student has clearly set out working in the box and both derivatives are correct.

↑ The correct derivatives from part (a) are equated and the resulting linear equation is solved to obtain the answer $x = -\frac{1}{2}$.

↑ Using a ruler to draw the tangent to the curve at $x = -\frac{1}{2}$ means that it is drawn accurately. The student has also checked the answer to part (c) by substituting $x = -\frac{1}{2}$ into the two derivative functions to see that the gradients of the line and tangent to the curve are the same.

Working:

(a) $f_1'(x) = 1$
$f_2'(x) = -2x$

(b) $1 = -2x$
$\frac{1}{-2} = x$

(c) $f_1'(-\frac{1}{2}) = f_2'(-\frac{1}{2}) = 1$

Answers:

(a) (i) $f_1'(x) = 1$ A1

(ii) $f_2'(x) = -2x$ A1 A1

(b) $-\frac{1}{2}$ A2

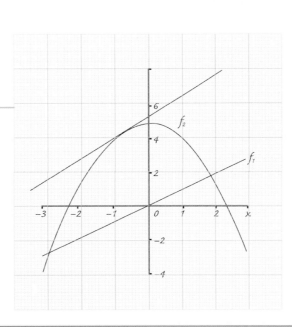

A1

Examiner report

Make sure that you can use the rule to differentiate the different terms of a polynomial, including a constant term. Remember that the derivative of a constant term is zero.

If you are asked to find the value of x where two graphs have the same gradient, you need to differentiate the two functions and set them equal to each other. Since part (a) asked for the derivatives of $f_1(x)$ and $f_2(x)$, part (b) involved solving $f_1'(x) = f_2'(x)$. Follow-through marks were awarded in part (b) if incorrect derivatives were found in part (a).

In part (c) students needed to remember that, if the gradient of the line and the gradient of the tangent of the curve are equal, then they will be parallel, since parallel lines have the same gradient. This question could have been answered without answering parts (a) or (b).

Common mistakes that students made in this question include not differentiating correctly, not appreciating that, if the gradient of two lines or tangents are equal, their gradients are the same, and forgetting to answer part of the question.

15. A function is represented by the equation

$$f(x) = ax^2 + \frac{4}{x} - 3.$$

(a) Find $f'(x)$. [3 marks]

The function $f(x)$ has a local maximum at the point where $x = -1$.

(b) Find the value of a. [3 marks]

[Taken from paper 1 November 2007]

How do I approach the question?

The information booklet gives the rule for finding the derivative: if $f(x) = ax^n$ then $f'(x) = nax^{n-1}$.

Rewriting the given function as $f(x) = ax^2 + 4x^{-1} - 3$ makes finding the derivative using the rule easier. Remember that the derivative of a constant term equals zero, so the derivative of the last term -3 will be zero.

Local maximum and minimum points occur when the tangent to the curve is horizontal, and at these points the gradient equals zero.

What is the correct notation?

If you are given the function as $f(x)$, the correct form of notation for the derivative is $f'(x)$.

This answer achieved 2/6

The first term is differentiated correctly, but the derivative of $\frac{4}{x}$ is incorrect, possibly due to too many steps being attempted together. The working out for differentiating this term could include the following steps:
$\frac{4}{x} = 4x^{-1} \Rightarrow \frac{d}{dx}(4x^{-1}) = -4x^{-2} = \frac{-4}{x^2}$
The lack of a third term in the derivative indicated that the student differentiated -3 correctly to get zero. The answer to this question was awarded 2 marks of the possible 3 marks.

The student has not realized that finding $f(-1)$ locates the y coordinate of the graph when $x = -1$. They have mistakenly assumed that this substitution will enable them to find the gradient. No marks are awarded in part (b), as the student has not shown they understand that, at $x = -1$, the gradient of the curve will be zero. Another way of writing this is $f'(-1) = 0$.

Working:

$f'(x) = 2ax$

$\rightarrow ?$
$a = -7$

$f(-1) = a - 1^2 + \frac{4}{-1} - 3$
$f(-1) = -a^2$ M0
$f(-1) = a = -7$ M0

Answers:

(a) $f'(x) = 2ax + \frac{8}{x^2}$ A2 A0

(b) $a = -7$ A0

10. Introductory differential calculus

This answer achieved 4/6

↑ The student has forgotten that the derivative of a constant term is zero and has just written down the -3 again without differentiating it. Two answer marks were awarded for correctly differentiating the first two terms and A0 for incorrectly differentiating the final constant term.

↑ A method mark has been awarded for setting the derivative equal to zero, in their incorrect equation. Because working is shown, it is clear that the student is using the correct method. A second follow-through mark is given for the correct substitution of $x = -1$ in the incorrect equation.

↓ Unfortunately, there is a mistake in the final line of working. The student's answer is clearly incorrect, because $2a = 7$ does not lead to $a = 3.8$; it should have been $a = 3.5$. This has then been changed to $a = 7$ in the answer line. Since $a = 7$ is incorrect, the examiner would look at the student's working and award marks for correct method. The student has provided two alternative values for a, but only one answer is required. The final answer mark is not awarded, as the question says find the value of a. Always check the wording carefully to see if one or more than one answer is required for a particular question.

Working:

$f(x) = ax^2 + \dfrac{4}{x} - 3$

$f(x) = ax^2 + 4x^{-1} - 3$

$f'(x) = 2ax - 4x^{-2} - 3$

$\qquad 2ax - \dfrac{4}{x^2} - 3$ A1 A1 A0

$2a(-1) - 4(-1)^{-2} - 3$

$0 = 2a(-1) - 4(-1)^{-2} - 3$ M1 M1(ft)

$0 = 2a(-1) - 4 - 3$

$-7 = 2a \times -1$

$7 = 2a$

$a = 3.8$

Answers:

(a) $f'(x) = 2ax - \dfrac{4}{x^2} - 3$

(b) $a = 7$ A0

This answer achieved 5/6

The student correctly differentiated the three terms to achieve full marks for part (a). Although there were no extra marks awarded for it, the term $-4x^{-2}$ was correctly rewritten as $-\frac{4}{x^2}$, which may have made the substitution of -1 in part (b) simpler, if it had been used in this way.

There were a few steps required in doing this question. First, the function $f(x)$ had a local maximum at $x = -1$. In mathematical terms, this meant that $f'(-1) = 0$. This student attempted too many algebraic steps at once. The student knew what had to be done, but made a mistake calculating $-4 \times (-1)^{-2}$. It would have been simpler to calculate $\frac{-4}{(-1)^2}$, especially since this form had been found in part (a). Since the method was correct and working was shown, the student was awarded 2 marks out of the possible 3 marks for part (b).

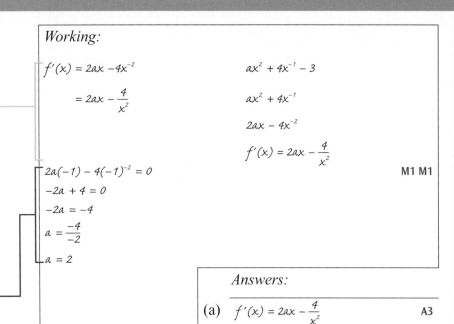

Working:

$f'(x) = 2ax - 4x^{-2}$

$\qquad = 2ax - \dfrac{4}{x^2}$

$ax^2 + 4x^{-1} - 3$

$ax^2 + 4x^{-1}$

$2ax - 4x^{-2}$

$f'(x) = 2ax - \dfrac{4}{x^2}$ \hfill M1 M1

$2a(-1) - 4(-1)^{-2} = 0$

$-2a + 4 = 0$

$-2a = -4$

$a = \dfrac{-4}{-2}$

$a = 2$

Answers:

(a) $f'(x) = 2ax - \dfrac{4}{x^2}$ \hfill A3

(b) $a = 2$ \hfill A0

Examiner report

Make sure you can find the derivative of a term with a negative index. Remember your rules for multiplying integers. A negative multiplied by another negative gives a positive number, and a negative multiplied by a positive gives a negative number. A negative number squared becomes positive. In this question, many students were able to find the derivative of the function but unable to correctly substitute $x = -1$ to find the gradient of the curve at that point.

Try not to skip steps when substituting values, as this can sometimes lead to errors being made. In question 15(b), writing down the steps where $f'(-1)$ is calculated does not take too much longer, but may result in a correct answer and therefore more marks being awarded for the question.

You must also appreciate that, if a particular x value has a gradient of zero, it means the tangent at that point is a horizontal line.

If the gradient to the left of a point with zero gradient is negative and that to the right is positive, the point of zero gradient represents a local **minimum** point. If the gradient to the left of the point with zero gradient is positive and that to the right is negative, the point is a local **maximum** point.

- Local minimum point: gradient changes from negative to positive.

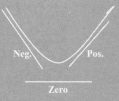

- Local maximum point: gradient changes from positive to negative.

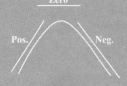

5. *[Maximum mark: 18]*

The diagram below shows the graph of a line L passing through $(1, 1)$ and $(2, 3)$ and the graph P of the function $f(x) = x^2 - 3x - 4$

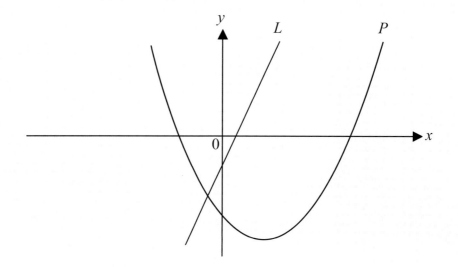

(a) Find the gradient of the line L. *[2 marks]*

(b) Differentiate $f(x)$. *[2 marks]*

(c) Find the coordinates of the point where the tangent to P is parallel to the line L. *[3 marks]*

(d) Find the coordinates of the point where the tangent to P is perpendicular to the line L. *[4 marks]*

(e) Find

   (i) the gradient of the tangent to P at the point with coordinates $(2, -6)$.

   (ii) the equation of the tangent to P at this point. *[3 marks]*

(f) State the equation of the axis of symmetry of P. *[1 mark]*

(g) Find the coordinates of the vertex of P and state the gradient of the curve at this point. *[3 marks]*

[Taken from paper 2 November 2007]

How do I approach the question?

Consider the graph and the information on the graph very thoroughly before starting. Mark the points (1, 1) and (2, 3) on the graph. Check what the various parts of the question require: gradients of lines and curves, gradients of perpendicular lines, equations of tangents, axis of symmetry of a parabola, and vertex of a parabola.

(a) The question relates to the line that passes through the two given points. This is a straightforward question and requires you to find the gradient of the line joining the two points. Remember that there is usually an increase in difficulty throughout a question, so parts (a) and (b) in each question in paper 2 should be reasonably simple.

(b) This asks you to differentiate the curve with equation $f(x) = x^2 - 3x - 4$. This derivative will be required later in the question.

The later parts of the question require you to analyse the information given and work out the mathematics required.

(c) When you are asked to find the coordinates of the point where the tangent to P is parallel to the line L, you first need to recall that parallel lines have the same gradient. You will then be using the information from parts (a) and (b) to answer the question. This happens quite often in paper 2 questions. So if you are having trouble getting started on a question, look at the previous parts to see if they can be used to help you solve a later part.

(d) The question asked for the point where the gradient was perpendicular. It is important to realize that, for perpendicular lines, the product of their gradients is equal to -1. If m_1 is the gradient of the first line and m_2 is the gradient of the perpendicular line, then $m_1 \times m_2 = -1$.

(e) Theory relating to the gradient and equation of the tangents to the curve is also being assessed in this part. The difference here is that another point is chosen on the curve, and this time you have to find $f'(2)$, recognize that this represents the gradient of the tangent at $(2, -6)$, and then find the equation.

The remaining parts (f) and (g) relate to the parabola P, and these questions may be answered using calculus, functions theory or your GDC.

This answer achieved 7/18

 The student has correctly calculated the gradient.

 Two of the three terms of the function have been correctly differentiated but the derivative of the constant term is incorrect. The student has forgotten that the derivative of a constant term is zero.

 The student did not see the relationship between the derivative function and the gradient of the line.

 A line perpendicular to line L, which has a gradient of 2, will have a gradient of $-\frac{1}{2}$.

 The student has correctly identified the axis of symmetry as $x = -\frac{b}{2a}$. From the axis of symmetry, the x coordinate of the vertex of the parabola is easily found and parts (f) and (g) are awarded full marks.

Working:

5 (a) $(1, 1)$ $(2, 3)$

$\dfrac{y_2 - y_1}{x_2 - x_1}$ $\dfrac{3 - 1}{2 - 1}$ $\dfrac{2}{1}$

$m = 2$ **M1 A1**

(b) $f(x) = x^2 - 3x - 4$

$= 2x - 3 - 4$ **A1**

(c) $y = mx + c$ $y = 2x + b$ **M0 A0**

(d) $y = mx + c$ $y = -2x + b$ **M0 A0**

(e) (i) -2 **A0**

(ii)

(f) $x = 1.4999...$ *(minimum value)* $x = -\dfrac{b}{2a}$

$x = 1.5$ **A1**

(g) $x = 1.5,$ $y = -6.25$ **G2**

$m = 0$ **A1**

This answer achieved 13/18

Parts (a) and (b) have been answered correctly.

Marks awarded were for finding the correct x coordinate of point P, but the final mark is lost due to the incorrect substitution into $f'(x)$ rather than correctly into $f(x)$.

The gradient and equation of the tangent at the point $(2, -6)$ were correctly found.

Writing "$= 1.5$" does not qualify as an equation. The student needed to write $x = 1.5$ to be awarded the 1 mark for this question.

The student has correctly identified the gradient of the curve at the vertex as zero, although the coordinates of the vertex are incorrect. Once again, the student has substituted a correct x coordinate into the wrong equation. The student has received the answer mark for stating that the gradient of curve at the vertex is zero.

Working:

5. (a) $\dfrac{3-1}{2-1} = \dfrac{2}{1}$ — M1

    $= 2$ — A1

(b) $f(x) = x^2 - 3x - 4$

    $f'(x) = 2x - 3$ — A2

(c) $2x - 3 = 2$ $\therefore 2\left(\dfrac{5}{2}\right) - 3 = 2$ — M1

    $2x = 5$ $\therefore (2.5, 2)$ — A1

    $x = \dfrac{5}{2}$ — A0

(d) $2x - 3 = -\dfrac{1}{2}$ $\therefore$ when $x = 1.25$ — M1 M1

    $2x = 2.5$ $2(1.25) - 3 = -0.5$

    $x = 1.25$ $\therefore (1.25, -0.5)$ — A1 A0

(e) (i) $x = 2, y = -6$

    when $x = 2$

    $2(2) - 3 = 1$ — A1

    (ii) $\therefore y + 6 = 1(x - 2)$ — M1

    $y + 6 = x - 2$

    $y = x - 8$ — A1

(f) $\dfrac{-b}{2a}$

    $= \dfrac{3}{2}$

    $= 1.5$ — A0

(g) $2(1.5) - 3 = 0$ — M0

    vertex $= (1.5, 0)$ — A0

    $\therefore$ gradient $= 0$ — A1

10. Introductory differential calculus

This answer achieved 18/18

The student has correctly equated the derivative of $f(x)$ with the gradient of line L. This has resulted in the equation $2x - 3 = 2$ and the correct answer that $x = 2.5$. Comments on the answer script such as "parallel lines have equal gradients" reveal an understanding of the background theory and the thought processes involved in each step.

The equation of the axis of symmetry has been found using the rule $x = -\frac{b}{2a}$.
Then $x = \frac{3}{2}$ is substituted into the original equation to find the coordinates of the vertex of the parabola. The coordinates of the vertex could also have been found directly from the GDC.

Working:

5 (a) $\frac{3-1}{2-1} = \frac{2}{1}$ Gradient of line $L = 2$ **M1 A1**

(b) $f(x) = x^2 - 3x - 4$
$f'(x) = 2x - 3$ **A2**

(c) $f'(x) =$ gradient of L
(parallel lines have = gradients)
$2x - 3 = 2$ **M1**
$2x = 5$
$x = 2.5$
$f(2.5) = (2.5)^2 - 3(2.5) - 4$
$= 6.25 - 7.5 - 4$ **A2**
$= -5.25$
Parallel to L at $(2.5, -5)$

(d) Perpendicular line will have a negative gradient (of L) $\Rightarrow \frac{-1}{2}$
$f'(x) = \frac{-1}{2}$ **M1**
$2x - 3 = \frac{-1}{2}$
$2x = 2.5$
$x = \frac{5}{4}$ **M1 A1**

$f\left(\frac{5}{4}\right) = \left(\frac{5}{4}\right)^2 - 3\left(\frac{5}{4}\right) - 4$
$= \frac{25}{16} - \frac{15}{4} - 4$
$= \frac{99}{16}$

Tangent to P = perpendicular to L at $\left(\frac{5}{4}, \frac{-99}{16}\right) \Rightarrow (1.25, -6.19)$ **A1**

(e) (i) $f'(2) = 2 \times 2 - 3$
$= 4 - 3$ **A1**
$= 1$

(ii) $y = mx + c, m = 1$
$-6 = 1 \times 2 + c$ **M1**
$c = -8 \Rightarrow y = x - 8$ **A1**

(f) $x = \frac{-b}{2a} = \frac{3}{2 \times 1} = \frac{3}{2} \rightarrow x = \frac{3}{2}$ **A1**

(1.5, −6.25)   **G2**

(g) $f\left(\frac{3}{2}\right) = \left(\frac{3}{2}\right)^2 - 3\left(\frac{3}{2}\right) - 4$ vertex $= \left(\frac{3}{2}, \frac{-25}{6}\right)$
$= \frac{-25}{6}$
Gradient $= 0 \rightarrow$ minimum   **A1**

120

Examiner report

This question required recall of rules for finding gradients and derivatives, and also some analysis to determine the mathematics required to answer the later parts of the question. The topic is about finding the gradients of tangents to curves and also the equations of tangents. It is important to review the concepts of linear equations when revising for this topic.

When asked to find *coordinates*, both x and y values are required. Many students have trouble finding a perpendicular gradient. If the gradient of L is 2, the gradient of the line perpendicular to L is $-\frac{1}{2}$, not -2.

In part (e), the question asked you to find the gradient of the tangent at the point $(2, -6)$. Once you have a point and a gradient, the equation of a line is easily found using a rule such as $y = mx + c$

The last two parts of the question could have been solved in several different ways. Marks are awarded for correct answers, irrespective of the method. It is important, if you are using your GDC, that you indicate how you are using it to solve the problem. Some alternative methods for these two parts are:

- to use calculus and find the x value where the gradient is zero. This would locate the axis of symmetry, and the y coordinate of the turning point could also be found.
- to use algebra and the equation $x = -\frac{b}{2a}$ to find the axis of symmetry. This x value is then used to substitute into $f(x)$ to find the y coordinate of the vertex (turning point).
- to use your GDC and sketch the graph. The axis of symmetry and vertex can then be read directly from the graph or found using the calculator.

Differentiation of the general term and the equation of the axis of symmetry can be found on pages 4 and 2, respectively, of the information booklet. When finding a particular x value, it is necessary to understand what you are going to use it for. If it is to find the gradient at a point, then the x value must be substituted into $f'(x)$. If it is to find a y coordinate of a point, then the x value must be substituted into $f(x)$.

If this is the last question on the exam paper, it is possible that you may be running out of time. It is very important to manage your time effectively, so that you have time to finish the last question and even check over your work.

11. Financial mathematics

Currency conversion

You should be able to:
- convert an amount of money given in specific currency units into another
- convert money in different currency units if a bank charges a commission
- evaluate different currency conversion options given different exchange rates.

You should know:
- each currency unit has a specific symbol, and all of the currency units have currency codes—for example, the euro has a currency sign €, and a currency code **EUR**.

Example
Convert 300 USD into Mexican pesos (MXN) if 1 USD buys 10.46 MXN, and the bank charges 1.3% commission.

Commission: 300 × 1.3% = 300 × 0.013 = 3.90 USD

Converted amount in (MXN): 296.1 × 10.46 = 3097.206 = 3097.21 MXN

Be prepared
- The commission rate is usually given as a percentage, so remember to convert it into decimals when used.

Simple interest

You should be able to:
- calculate one of the variables in the simple interest formula if the other three are known—for example, calculate the interest if the capital, interest rate and time are given
- calculate periodic repayments for a loan.

You should know:
- simple interest formula
$$I = \frac{Crn}{100}$$
where C = capital, $r\%$ = interest rate, n = number of time periods, and I = interest

Example
Calculate the monthly repayments on a loan of 3070 Argentine pesos (ARS) borrowed at 7.5% p.a. simple interest over five years.

If the capital amount is 3070 ARS, then the interest is

$$I = \frac{3070 \times 7.5 \times 5}{100} = 1151.25 \, ARS$$

Thus, the amount that should be repaid is 3070 + 1151.25 = 4221.25 ARS. Since we want to calculate the monthly repayments, this would be repaid over a period of 5 × 12 = 60 months. Therefore, the monthly repayment is

$$\frac{4221.25}{60} = 70.35 \, ARS$$

where the answer is given correct to two decimal places.

Be prepared
- Keep in mind exactly what you are asked to calculate. For example, if it is the interest accumulated by a certain amount invested in a bank, then you calculate it by using the formula above. If you are asked to find the entire amount of money you have in a bank after you have left it for a certain period of time, then you have to add the interest to the amount originally invested.
- It is usually more advantageous in terms of saving time to use your GDC in answering the types of questions discussed above.

Compound interest

You should be able to:

- calculate one of the variables in the compound interest formula if the other four are known—for example, calculate the interest given the capital, nominal interest rate, time and number of compounding periods in one year
- compare different investment options
- calculate depreciated values.

You should know:

- compound interest formula

$$I = C \times \left(1 + \frac{r}{100k}\right)^{kn} - C$$

where C = capital, $r\%$ = nominal interest rate, n = number of years, k = number of compounding periods in one year, and I = interest

- total accumulated amount

$$T = C \times \left(1 + \frac{r}{100k}\right)^{kn}$$

Example

Sonya has to decide whether to invest 5000 Bulgarian levs (BGN) for three years in Unisbank or Bulgarbank. At the moment, Unisbank offers a nominal interest rate of 5.05% compounded monthly on all savings accounts, and Bulgarbank offers a nominal interest rate of 5.8% compounded annually. Calculate the amount she would accumulate in either case, and decide which is the better choice for her.

Total amount to be accumulated in Unisbank is
$$T_1 = 5000 \times \left(1 + \frac{5.05}{100 \times 12}\right)^{36} = 5816.04 \, BGN$$

Total amount to be accumulated in Bulgarbank is
$$T_2 = 5000 \times \left(1 + \frac{5.8}{100}\right)^{3} = 5921.44 \, BGN$$

From the above calculations we can conclude that Bulgarbank would be a better investment choice.

Texas Instruments

APPS – Finance

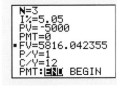 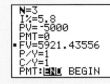

Casio

MENU – TVM

Be prepared

- Choose the appropriate formula carefully—whether the interest or the total accumulated amount formula—if you need one.

Construction and use of tables: loan and repayment schemes, investment and saving schemes

You should be able to:

- read, interpret and use tables that represent loan and repayment schemes
- use the tables to compare two different schemes for repayment of a loan or savings.

Example

Sasha takes out a loan of $14 250 and negotiates an offer from the bank for three years at 7.55% p.a. interest. Use the circled entry in the table to calculate the following amounts.

Table of monthly repayments per $1000					
	Annual interest rate				
Loan terms (months)	7.05%	7.30%	7.55%	8.05%	7.30%
6	172.4416	172.6429	172.8439	173.2452	173.4455
12	89.2083	89.4167	89.625	90.0417	90.25
24	47.7488	47.9720	48.1958	48.6450	48.8704
30	39.5226	39.7537	39.9857	40.4521	40.6865
36	34.0767	34.3160	(34.5564)	35.0406	35.2844

(a) The monthly repayments.

Information from the table tells us that $34.5564 is the monthly repayment for every $1000 borrowed for three years (36 months) at 7.55% interest. Since 14 250 : 1000 = 14.25, then the monthly repayment is
14.25 × 34.5564 = 492.4287 = $492.43

(b) The total repayments.

This is 36 times the monthly repayment, that is
36 × 492.43 = $17 727.48

(c) The interest that he has to pay.

$17 727.48 − $14 250 = $3477.48

Be prepared

- Usually when tables are given, you **must** use them to answer the questions. Often there is not specific information about how the interest is calculated. For example, if there is no information about the number of compounding periods, then you cannot use the interest formulae given on the previous page.

5. The table below shows the **monthly** repayments per $10 000 borrowed for various nominal annual interest rates.

Loan term (years)	Table of monthly repayments in $, **per $10 000** Annual interest rate		
	7%	8%	9%
5	198.0112	202.7634	207.5836
10	116.1085	121.3276	126.6758
15	89.8828	95.5652	101.4267
20	77.5299	83.6440	89.9726
25	70.6779	77.1816	83.9196

Beryl borrows $150 000 to buy an apartment at an interest rate of 8%, to be repaid over 20 years.

(a) Calculate Beryl's exact monthly repayment.

(b) Find the exact amount of **interest** paid for the loan over the 20 years.

[Taken from paper 1 May 2007]

How do I approach this question?

All the information is provided within the question. You just have to take this and work out the answer by reading the table. You should also be aware of the key words "monthly" and "per $10 000" in bold. You will need to understand clearly what the term "monthly repayment" means. It is the total amount (that is, the amount borrowed plus the interest charged) divided by the number of months in the loan period.

(a) Within the table, you are able to locate the monthly repayment for a loan of $10 000 at an 8% interest rate over 20 years. You are told how large Beryl's loan is. When you have both of these, it is not difficult to find the answer to part (a) of the question.

(b) Again, the word "interest" is in bold here to help to keep you focused. The answer to part (a) will allow you to work out the total amount that Beryl has to repay. After this, all you need to do is to make one final step to work out the interest to answer part (b) of the question.

What are the key areas from the syllabus?

- Interest
- Loans
- Use of tables for repayment schemes

This answer achieved 0/6

Here, the student has made two mistakes. First, 20 is not the number of months in a 20-year period. It should be $20 \times 12 = 240$ months. Second, the amount 83.6440 used from the table is not the monthly repayment of the interest alone, but also includes the repayment on the amount borrowed. The product 83.6440×240 represents the total amount that has to be repaid. To find out only the interest that is paid on the loan, the borrowed amount $150\,000 has to be subtracted from the total amount repaid.

The student has forgotten that the loan needs to be paid with interest. In order to find out the monthly repayments, the student should use the table given.

Working:

(a) $\dfrac{\$150{,}000}{240 \text{ months}} = 625$

(b) $83{,}6440(20) = 1672.88$

Answers:

(a) $625   A0 M0 A0

(b) 1672.88   M0 A0 A0

This answer achieved 4/6

A correct and clear solution is presented. The student has chosen the correct value from the table.

The student assumes incorrectly that the total repayment is calculated by the use of the simple interest formula. The question does not specify whether the interest is calculated as a simple or compound interest, so we cannot assume that. The student should have made a connection between parts (a) and (b) and used the monthly repayment calculated in (a) to calculate the total repayment.

The calculated total repayment is incorrect, but 1 mark is still awarded to part (b) as the last line in the solution shows that the student understands that the interest is found by subtracting the borrowed amount from the total repayment.

Working:

(a) $\dfrac{150\,000}{10\,000} = 15$

$83.6440 \times 15 = 1254.66$

(b) $I = \dfrac{150\,000 \times 8 \times 20}{100}$

$= 240\,000$ M0 A0

$240\,000 - 150\,000 = 90\,000$

Answers:

(a) 1254.66 $ C3

(b) 90 000 $ A1ft

This answer achieved 6/6

The number of months for repayment of the loan are calculated. Then, using the answer from (a), the student calculates the total amount to be paid. To achieve the correct answer, from the total amount the student subtracts the amount that was borrowed, to find the interest that will be paid.

Working:

$150{,}000 \div 10{,}000 = 15$

$83.6440 \times 15 = 1254.66$

$12 \times 20 = 240$ *total months paid*

$240 \times 1254.66 = \$301{,}118.40$ *paid over 20 years*

$301{,}118.40 - 150{,}000 = \$15{,}118.40$

Answers:

(a) $1254.66   C3

(b) $151,118.40   C3

Both answers given to two decimal places.

Examiner report

A common mistake with this type of question is that students have difficulty in reading and interpreting tables that represent loan repayments. Mistakes are made as a result of misreading the table, and not keeping in mind both what is given and what is asked for. In this question, the table refers to loans of $10 000 for different loan periods, and different interest rates. So students need to be able to select the monthly repayment after having clearly identified the given terms of the loan.

Often students do not remember that they are dealing with a 15 times bigger loan, and end up using in their calculations the monthly repayment for a $10 000 loan.

Another mistake that is made here is attempting to calculate the amount of the total loan repaid by multiplying the monthly repayment by the number of the years of the loan, instead of by the number of months.

Before starting work on part (b), check whether the answer to (a) can be used in solving (b), and, if so, how. Often students try to use one of the formulae for calculating the interest on a loan. However, this is not possible here, as there is not enough information. It is also necessary to know what the relation is between the interest on the loan, the amount of the loan, and the total amount to be repaid.

6. Two brothers Adam and Ben each inherit $6500. Adam invests his money in a bond that pays simple interest at a rate of 5% per annum. Ben invests his money in a bank that pays compound interest at a rate of 4.5% per annum.

(a) Calculate the value of **Adam's** investment at the end of 6 years. *[3 marks]*

(b) Calculate the value of **Ben's** investment at the end of 6 years. Give your answer **correct to 2 decimal places**. *[3 marks]*

[Taken from paper 1 November 2007]

How do I approach the question?

After reading the question you should be aware that it is about two different investment schemes: (a) is a simple interest, and (b) is a compound interest scheme. Since you have to calculate the amount that is accumulated over six years in the two cases, you need to use the formulae for simple interest and for compound interest, and then find the total amounts. You are encouraged to use your GDC to calculate the amounts that both investment schemes yield. The TI and Casio screenshots from the calculation of the value of Ben's investment are shown below.

Texas Instruments

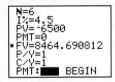

Casio

How does this relate to the information formula booklet?

Carefully evaluate the information given in the question in order to decide whether you need to use the simple interest formula or the compound interest formula. Also, be careful about what you have to calculate—only the interest on the invested amount of money or the total amount accumulated over six years.

What are the key areas from the syllabus?

- Simple interest
- Compound interest

This answer achieved 2/6

Working:

```
    A              B
   6500           6500
    ↓              ↓
  5 5% pa       C 4.5% pa
    ↓              ↓
  6 years       6 years
      ↘           ↓
              72 months
```

The student uses the simple interest formula, which is the correct choice. However, the substitution in the formula is incorrect, as the time is given as 72 years instead of 6.

(a) $6 \times 12 = 72$

$I = \dfrac{Cm}{100}$

$I = \dfrac{6500 \times 5 \times 72}{100}$ M1 A0

$I = \$23400$ A0

The student correctly identifies the compound interest formula to be used for calculating the interest. Although, the choice of the formula is correct, the substitution and final answer are incorrect.

(b) $I = 6500\left(1 + \dfrac{4.5}{100}\right) - 6500$ M1 A0

$I = 6500(23.78882066) - 6500$

$I = 148127.3343$ A0

150000

Only 1 mark is given for the correct method of calculating the interest.

Answers:

(a) $23400.00

(b) $150000.00

This answer achieved 4/6

The student chooses the correct simple interest formula and makes a correct substitution. As a result, the obtained interest is correct, and the student is awarded 1 mark for correct method, and an additional mark for correct calculation of the interest.

The compound interest is calculated accurately, and 2 marks are awarded for correct method and correct calculation of the interest.

The student ends their solution early and does not calculate the total amount, which results in 1 mark lost.

An accuracy penalty is incurred because of inaccurate rounding.

Working:

(a) $I = \dfrac{6500 \times 5 \times 6}{100}$ M1 A1

$I = \$1950$

(b) $= 8464.6908$ M1 A1

$= 8464.7$

Answers:

(a) $1950 A0

(b) $8464.70 A0 AP

This answer achieved 6/6

The student chooses the appropriate formula for calculation of the simple interest and receives 1 mark for it. It is then used correctly to calculate the simple interest, and the student receives another mark for it. Finally, the calculated interest is added to the capital amount in order to calculate the total amount, for which a third mark is awarded.

Full marks are achieved: 1 mark is awarded for a correct choice of the compound interest formula, another is given for correct substitution, and the final mark is given for the correct answer appropriately rounded to two decimal places.

Working:

$i = \dfrac{6500 \times 5 \times 6}{100}$

$= 1950 + 6500$

$= \$8450$

$i = 6500 \left(1 + \dfrac{4.5}{100}\right)^6$

$= \$8464.69$

Answers:

(a) $8450.00 C3

(b) $8464.69 C3

Examiner report

In a financial question like this one, the first important steps are to identify the appropriate formula for calculation of the interest, and then to substitute in the formula with the correct values. Students seldom make mistakes in choosing the formula, but mistakes are often made during the substitution process, as in the first two examples. It may help if, before using a formula, you consult the booklet and remind yourself what each variable means. Also, very often students forget to add the initial capital invested to the interest when the question asks for the total amount of money accumulated over a time period.

(ii) Robert, who lives in the UK, travels to Belgium. The exchange rate is 1.37 euros to one British Pound (GBP) with a commission of 3 GBP, which is subtracted before the exchange takes place. Robert gives the bank 120 GBP.

(a) Calculate **correct to 2 decimal places** the amount of euros he receives. *[3 marks]*

He buys 1 kilogram of Belgian chocolates at 1.35 euros per 100 g.

(b) Calculate the cost of his chocolates in GBP **correct to 2 decimal places**. *[3 marks]*

[Taken from paper 2 November 2007]

How do I approach the question?

This is a mixed question, involving currency and SI unit conversion.

(a) For this part, identify the currency that you convert from and the currency you convert into. A simple chart like this one may help you with the conversion:

1 GBP = 1.37 EUR

117 GBP = ?? EUR

Of course, you need to pay the bank its commission, so you will end up 3 GBP short!

(b) One way to approach this part of the question is first to calculate how much the chocolate that Robert buys will cost in euros, and then to convert it into GBP. If you convert 1 kg into g, and find out that you need to buy 10 bars (100 g each) of chocolate, then your next step is to find their cost in euros. Finally, you convert the euros into GBP. Again, a simple chart like the following may help you in the conversion:

1.37 EUR = 1 GBP

1 EUR = ?? GBP

13.5 EUR = ?? GBP

What are the key areas from the syllabus?

- Currency conversion
- SI units conversion
- Units: kg and g

This answer achieved 2/6

This answer indicates a misconception about commission—the student adds instead of subtracts 3 GBP from the initial amount that Robert is converting.

The student uses a correct method for conversion from GBP to EUR.

The mark is given for a correct final answer only.

The student writes that 1 kg equals 10 000 g. Incorrect!

Another 0 mysteriously attaches itself to 10 000 g and turns it into 100 000 g. A third mistake with the multiplication follows. The student calculates that the cost of 1 kg of chocolate is 13 500 EUR. As a result, a kilogram of chocolate costs as much as a new car! You do not have to do mathematics to know that this cannot be correct. A mark is lost for the incorrect calculation of the cost of 1 kg of chocolate.

Working:

3 (ii) (a)   1.37€ = 1 GBP   3GBP   A0

   (a)   123GBP × 1.37 = 168.51   M1
         168.51€   A0

   (b)   1.35   1 kg = 10000 g   A0
         1.35 × 100000 = 13500
         $\frac{13500}{1.37} = 9854.01 GBP$ M1 A0

The student divides the cost of 1 kg of chocolate by 1.37 and thus achieves a method mark for using a correct method for conversion from euros to GBP. But the mistakes mentioned above prevent the student from acquiring the correct final answer, and the final mark is lost.

This answer achieved 3/6

The student has calculated the amount with the subtracted commission.

One mark is given for using a correct method for converting euros to GBP.

The student correctly starts off by converting the cost of 100 g of chocolate from euros to GBP. The method is appropriate, and 1 mark is awarded. However, the student does not complete the answer by multiplying by 10 to find the cost of 1 kg of chocolate.

Working:

(ii) (a)   120 GBP − 3
         = 117 GBP   A1
         → Convert to Euro
         = 117 GBP × 1.37   M1
         = 160.3 euros
         = 160 euros   A0

   (b)   = euros = $\frac{1.35}{1.37}$ = 0.986   M1 A0
                = 0.98 pence GBP at per 100 g   A0

This answer achieved 6/6

A correct solution with clear reasoning is presented here. One mark is given for subtracting the commission from Robert's initial amount. Another mark is awarded for a correct method chosen for conversion of the amount in euros into GBP. The final mark is given for achieving the correct final answer.

Working:

(ii) (a) 120 GBP − 3 = 117 GBP (after commission)   A1
   117 × 1.37 = 160.29€   M1
   ∴ he receives 160.29€ for 117 GBP   A1

(b) 1 kg = 10 × 100 g
   ∴ 1.35 × 10 = cost of 1 kg   A1
   = 13.5€   M1
   $\dfrac{13.5€}{1.37}$ = 9.85 GBP   A1

Mark awarded for finding the cost of 1 kg of chocolate.

Correct approach to converting euros to GBP.

Well done, full marks. The student finds the cost of 1 kg of chocolate by concluding that it will be 10 times more than the cost of 100 g, and thus 13.5€. The student then converts the cost calculated in euros into GBP. Note how clear the reasoning and the write-up of the entire response is.

Examiner report

A common difficulty here is determining how to convert from one currency unit into another. We often encounter students' misconceptions about commission, and what it implies. Another common mistake is related to converting SI units—in this case converting kilograms into grams. In summary, to be able to solve this problem, knowledge of currency conversion, SI unit conversion, and commission is necessary. What is equally important is to always try to make sense of the operations you perform! Check also whether your results are reasonable, and that you do not conclude that a chocolate bar costs as much as a car!

12. Are you ready?

We hope that you enjoyed the book and that you will continue to use it throughout the year to help you to prepare for your exams.

Finally, we have added a set of past exam papers from May 2008 to which you may apply your new skills. You can either take these under exam conditions or refer back to the book, the choice is yours.

Paper 1

1. (a) Calculate exactly $\dfrac{(3 \times 2.1)^3}{7 \times 1.2}$. *[1 mark]*

   (b) Write the answer to part (a) correct to 2 significant figures. *[1 mark]*

   (c) Calculate the percentage error when the answer to part (a) is written correct to 2 significant figures. *[2 marks]*

   (d) Write your answer to **part (c)** in the form $a \times 10^k$ where $1 \leq a < 10$ and $k \in \mathbb{Z}$. *[2 marks]*

2. There are 120 teachers in a school. Their ages are represented by the cumulative frequency graph below.

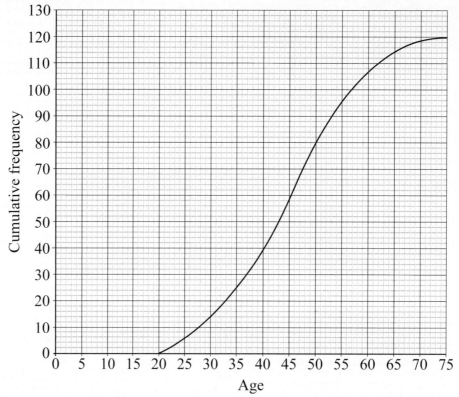

(a) Write down the median age. [1 mark]

(b) Find the interquartile range for the ages. [2 marks]

(c) Given that the youngest teacher is 21 years old and the oldest is 72 years old, represent the information on a box and whisker plot using the scale below. [3 marks]

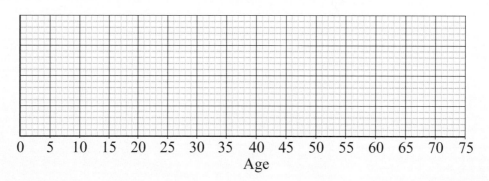

3. Consider the function $f(x) = 2x^3 - 5x^2 + 3x + 1$.

   (a) Find $f'(x)$. *[3 marks]*

   (b) Write down the value of $f'(2)$. *[1 mark]*

   (c) Find the equation of the tangent to the curve of $y = f(x)$ at the point $(2,3)$. *[2 marks]*

4. (a) Complete the following table of values for the height and weight of seven students. *[4 marks]*

	Values	Mode	Median	Mean	Standard deviation
Height (cm)	151, 158, 171, 163, 184, 148, 171			164	11.7
Weight (kg)	53, 61, 58, 82, 45, 72, 82	82	61		

The ages (in months) of seven students are 194, 205, 208, 210, 200, 226, 223.

   (b) Represent these values in an ordered stem and leaf diagram. *[2 marks]*

5. 200 people of different ages were asked to choose their favourite type of music from the choices Popular, Country and Western and Heavy Metal. The results are shown in the table below.

Age/Music choice	Popular	Country and Western	Heavy Metal	Totals
11 – 25	35	5	50	90
26 – 40	30	10	20	60
41 – 60	20	25	5	50
Totals	85	40	75	200

It was decided to perform a chi-squared test for independence at the 5 % level on the data.

(a) Write down the null hypothesis. *[1 mark]*

(b) Write down the number of degrees of freedom. *[1 mark]*

(c) Write down the chi-squared value. *[2 marks]*

(d) State whether or not you will reject the null hypothesis, giving a clear reason for your answer. *[2 marks]*

6. Consider the following logic propositions:

p: Sean is at school
q: Sean is playing a game on his computer.

(a) Write in words, $p \veebar q$. *[2 marks]*

(b) Write in words, the converse of $p \Rightarrow \neg q$. *[2 marks]*

(c) Complete the following truth table for $p \Rightarrow \neg q$. *[2 marks]*

p	q	$\neg q$	$p \Rightarrow \neg q$
T	T		
T	F		
F	T		
F	F		

7. Triangle ABC is such that AC is 7 cm, angle ABC is 65° and angle ACB is 30°.

(a) Sketch the triangle writing in the side length and angles. *[1 mark]*

(b) Calculate the length of AB. *[2 marks]*

(c) Find the area of triangle ABC. *[3 marks]*

8. The first term of an arithmetic sequence is 0 and the common difference is 12.

   (a) Find the value of the 96th term of the sequence. *[2 marks]*

   The first term of a geometric sequence is 6. The 6th term of the geometric sequence is equal to the 17th term of the arithmetic sequence given above.

   (b) Write down an equation using this information. *[2 marks]*

   (c) Calculate the common ratio of the geometric sequence. *[2 marks]*

---

9. The table below shows the number of words in the extended essays of an IB class.

Number of words	$3200 \leq w < 3400$	$3400 \leq w < 3600$	$3600 \leq w < 3800$	$3800 \leq w < 4000$	$4000 \leq w < 4200$
Frequency	2	5	8	17	3

   (a) Draw a histogram on the grid below for the data in this table. *[3 marks]*

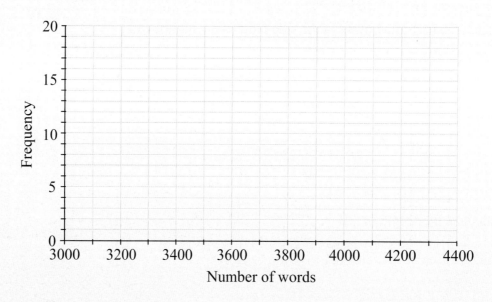

   (b) Write down the modal group. *[1 mark]*

   The maximum word count is 4000 words.

   (c) Write down the probability that a student chosen at random is on or over the word count. *[2 marks]*

10. Jane plans to travel from Amsterdam to Chicago. She changes 1500 Euros (EUR) to US Dollars (USD) at an exchange rate of 1 EUR to 1.33 USD. Give all answers in this question **correct to two decimal places**.

   (a) Calculate the number of USD Jane receives. *[1 mark]*

   Jane spends 1350 USD and then decides to convert the remainder back to EUR at a rate of 1 EUR to 1.38 USD.

   (b) Calculate the amount of EUR Jane receives. *[3 marks]*

   If Jane had waited until she returned to Amsterdam she could have changed her USD at a rate of 1 EUR to 1.36 USD but the bank would have charged 0.8 % commission.

   (c) Calculate the amount of EUR Jane gained or lost by changing her money in Chicago. *[2 marks]*

11. (a) Consider the numbers 2, $\sqrt{3}$, $-\dfrac{2}{3}$ and the sets $\mathbb{N}$, $\mathbb{Z}$, $\mathbb{Q}$ and $\mathbb{R}$.

   Complete the table below by placing a tick in the appropriate box if the number is an element of the set, and a cross if it is not.

		$\mathbb{N}$	$\mathbb{Z}$	$\mathbb{Q}$	$\mathbb{R}$
(i)	2				
(ii)	$\sqrt{3}$				
(iii)	$-\dfrac{2}{3}$				

   [3 marks]

   (b) A function f is given by $f : x \mapsto 2x^2 - 3x$, $x \in \{-2, 2, 3\}$.

   (i) Draw a mapping diagram to illustrate this function.

   (ii) Write down the range of function f. *[3 marks]*

12. The following table shows the monthly payments needed to repay a loan of $1000 with various rates and time periods.

Table of Monthly Repayments per $1000 Annual interest rate				
Loan Term (months)	**5 %**	**5.5 %**	**6 %**	**6.5 %**
12	87.50	87.92	88.34	88.75
18	59.74	60.21	60.62	61.12
24	45.94	46.38	46.84	47.25
30	37.66	38.11	38.57	39.04
36	32.16	32.62	33.09	33.56
42	28.25	28.72	29.20	29.70
48	25.33	25.81	26.30	26.80

Sarah takes out a personal loan for $24 000 to buy a car. She negotiates a loan for three years at 6 % per annum interest.

(a) Calculate the exact monthly repayment she will make. *[2 marks]*

(b) Find the exact total of the repayments she will make. *[2 marks]*

Beryl took out a loan of $10 000 for 18 months. The total she paid for the loan was $10 837.80.

(c) Find the rate of interest charged on the loan. *[2 marks]*

13. Bob invests 3000 USD in a bank that offers simple interest at a rate of 4 % per annum.

(a) Calculate the number of years that it takes for Bob's money to double. *[3 marks]*

Charles invests 3000 USD in a bank that offers compound interest at a rate of 3.5 % per annum, compounded half-yearly.

(b) Calculate the number of years that it takes for Charles's money to double. *[3 marks]*

14. A race track is made up of a rectangular shape 750 m by 500 m with semi-circles at each end as shown in the diagram.

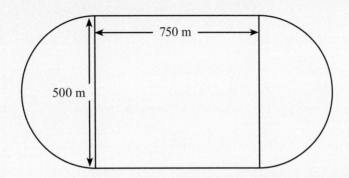

Michael drives around the track once at an average speed of 140 kmh⁻¹.

(a) Calculate the distance that Michael travels. [2 marks]

(b) Calculate how long Michael takes in **seconds**. [4 marks]

15. In an experiment it is found that a culture of bacteria triples in number every four hours. There are 200 bacteria at the start of the experiment.

Hours	0	4	8	12	16
No. of bacteria	200	600	a	5400	16200

(a) Find the value of a. [1 mark]

(b) Calculate how many bacteria there will be after one day. [2 marks]

(c) Find how long it will take for there to be two million bacteria. [3 marks]

Paper 2

1. *[Maximum mark: 17]*

(a) Sketch the graph of the function $f(x) = \dfrac{2x+3}{x+4}$, for $-10 \leq x \leq 10$. Indicating clearly the axis intercepts and any asymptotes. *[6 marks]*

(b) Write down the equation of the vertical asymptote. *[2 marks]*

(c) On the same diagram sketch the graph of $g(x) = x + 0.5$. *[2 marks]*

(d) Using your graphical display calculator write down the coordinates of **one** of the points of intersection on the graphs of f and g, **giving your answer correct to five decimal places**. *[3 marks]*

(e) Write down the gradient of the line $g(x) = x + 0.5$. *[1 mark]*

(f) The line L passes through the point with coordinates $(-2, -3)$ and is perpendicular to the line $g(x)$. Find the equation of L. *[3 marks]*

2. *[Maximum mark: 20]*

(i) A group of 50 students completed a questionnaire for a Mathematical Studies project. The following data was collected.

        18 students own a digital camera (D)
        15 students own an ipod (I)
        26 students own a cell phone (C)
        1 student owns all three items
        5 students own a digital camera and an ipod but not a cell phone
        2 students own a cell phone and an ipod but not a digital camera
        3 students own a cell phone and a digital camera but not an ipod

(a) Represent this information on a Venn diagram. *[4 marks]*

(b) Calculate the number of students who own none of the items mentioned above. *[2 marks]*

(c) If a student is chosen at random, write down the probability that the student owns a digital camera **only**. *[1 mark]*

(d) If two students are chosen at random, calculate the probability that they both own a cell phone **only**. *[3 marks]*

(e) If a student owns an ipod, write down the probability that the student also owns a digital camera. *[2 marks]*

(ii) Claire and Kate both wish to go to the cinema but one of them has to stay at home to baby-sit.
The probability that Kate goes to the cinema is 0.2. If Kate does not go Claire goes.
If Kate goes to the cinema the probability that she is late home is 0.3.
If Claire goes to the cinema the probability that she is late home is 0.6.

(a) Copy and complete the probability tree diagram below. *[3 marks]*

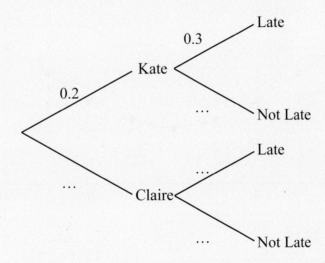

(b) Calculate the probability that

(i) Kate goes to the cinema and is not late; *[2 marks]*

(ii) the person who goes to the cinema arrives home late. *[3 marks]*

3. *[Maximum mark: 13]*

(i) The graph below represents the temperature ($T°$ Celsius) in Washington measured at midday during a period of thirteen consecutive days starting at Day 0. These points also lie on the graph of the function

$$T(x) = a + b\cos(cx°), \ 0 \le x \le 12,$$

where a, b and $c \in \mathbb{R}$.

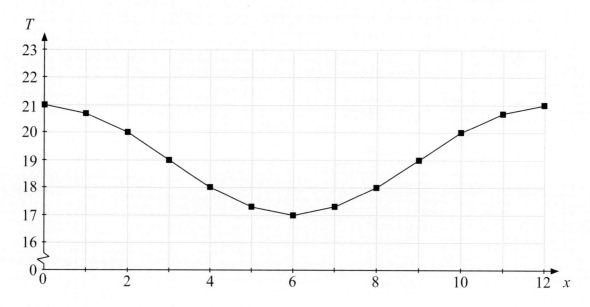

(a) Find the value of

   (i) a;

   (ii) b. *[2 marks]*

(b) Show that $c = 30$. *[1 mark]*

(c) Using the graph, or otherwise, write down the part of the domain for which the midday temperature is less than $18.5°$. *[2 marks]*

(ii) The number of bottles of water sold at a railway station on each day is given in the following table.

Day	0	1	2	3	4	5	6	7	8	9	10	11	12
Temperature ($T°$)	21	20.7	20	19	18	17.3	17	17.3	18	19	20	20.7	21
Number of bottles sold (n)	150	141	126	125	98	101	93	99	116	121	119	134	141

(a) Write down

  (i) the mean temperature;

  (ii) the standard deviation of the temperatures. *[2 marks]*

(b) Write down the correlation coefficient, r, for the variables n and T. *[1 mark]*

(c) Comment on your value for r. *[2 marks]*

(d) The equation of the line of regression for n on T is $n = dT - 100$.

  (i) Write down the value of d.

  (ii) Estimate how many bottles of water will be sold when the temperature is 19.6°. *[2 marks]*

(e) On a day when the temperature was 36° Peter calculates that 314 bottles would be sold. Give one reason why his answer might be unreliable. *[1 mark]*

4. *[Maximum mark: 19]*

   (i) Mal is shopping for a school trip. He buys 50 tins of beans and 20 packets of cereal. The total cost is 260 Australian dollars (AUD).

   (a) Write down an equation showing this information, taking b to be the cost of one tin of beans and c to be the cost of one packet of cereal in AUD. *[1 mark]*

   Stephen thinks that Mal has not bought enough so he buys 12 more tins of beans and 6 more packets of cereal. He pays 66 AUD.

   (b) Write down another equation to represent this information. *[1 mark]*

   (c) Find the cost of one tin of beans. *[2 marks]*

   (d) (i) Sketch the graphs of these two equations.

   (ii) Write down the coordinates of the point of intersection of the two graphs. *[4 marks]*

   (ii) The triangular faces of a square based pyramid, ABCDE, are all inclined at 70° to the base. The edges of the base ABCD are all 10 cm and M is the centre. G is the mid-point of CD.

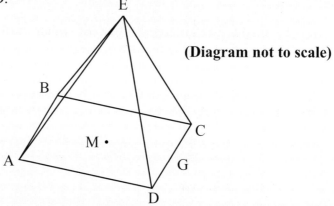

(Diagram not to scale)

   (a) Using the letters on the diagram draw a triangle showing the position of a 70° angle. *[1 mark]*

   (b) Show that the height of the pyramid is 13.7 cm, to 3 significant figures. *[2 marks]*

   (c) Calculate

   (i) the length of EG;

   (ii) the size of angle DEC. *[4 marks]*

   (d) Find the total surface area of the pyramid. *[2 marks]*

   (e) Find the volume of the pyramid. *[2 marks]*

5. *[Maximum mark: 21]*

(i) (a) Factorise $3x^2 + 13x - 10$. *[2 marks]*

(b) Solve the equation $3x^2 + 13x - 10 = 0$. *[2 marks]*

Consider a function $f(x) = 3x^2 + 13x - 10$.

(c) Find the equation of the axis of symmetry on the graph of this function. *[2 marks]*

(d) Calculate the minimum value of this function. *[2 marks]*

(ii) A closed rectangular box has a height y cm and width x cm. Its length is twice its width. It has a fixed outer surface area of 300 cm^2.

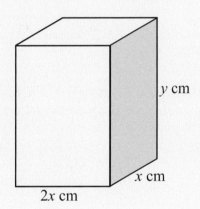

(a) Show that $4x^2 + 6xy = 300$. *[2 marks]*

(b) Find an expression for y in terms of x. *[2 marks]*

(c) Hence show that the volume V of the box is given by $V = 100x - \dfrac{4}{3}x^3$. *[2 marks]*

(d) Find $\dfrac{dV}{dx}$. *[2 marks]*

(e) (i) Hence find the value of x and of y required to make the volume of the box a maximum.

(ii) Calculate the maximum volume. *[5 marks]*

Notes